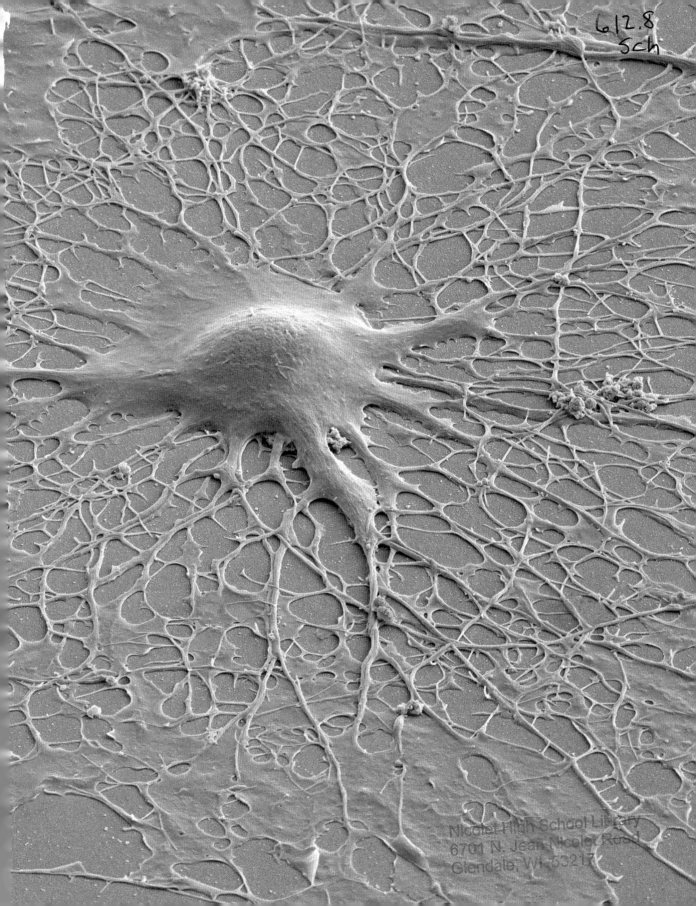

Abrams, New York

PORTRAITS
OF THE
MIND

Visualizing the Brain
from Antiquity to the 21st Century

CARL SCHOONOVER

Foreword by Jonah Lehrer

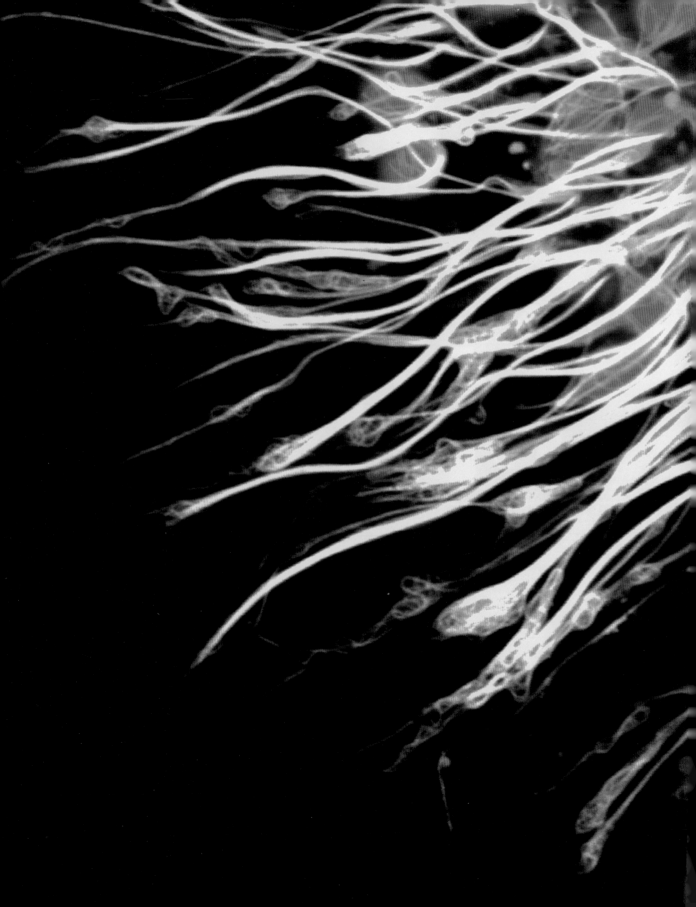

CONTENTS

FOREWORD
by Jonah Lehrer

The human brain is not a black box, and it never was. The first time I saw a naked cortex, freshly removed from its bony encasing, I was struck by its bloodiness. There was no soul here, just a thinking machine made of flesh and fat, dense with purple veins and leaking all sorts of spooky fluids. I couldn't believe that I had emerged from a similar mass, just these three pounds of meat with the texture of Jell-O.

This is the brain we see with our bare eyes. It's so brutishly of the body that it's not surprising that people assumed for thousands of years that there must be something else, some invisible substrate that explained the metaphysical aspects of the mind. It seems ridiculous, after all, that such a material organ could give rise to the experience of an emotion, or the taste of a peach, or the words in this sentence.

And yet, there is nothing else: This is all we are. The power of this beautiful book is to show us that the fleshy brain is more than enough, that it contains the multitudes and the machinery necessary to explain the wonder of our existence. From the Technicolor looms of the Brainbow to the sprawl of the single spiny neuron glimpsed by an electron microscope to the vast axonal networks revealed by diffusion magnetic resonance imaging—it's clear that beneath the convoluted surface of the cortex is a biological design of near infinite intricacy. The soul isn't dead, it doesn't seem to be needed.

It's worth taking a moment to appreciate the newness of these brain pictures. For thousands of years, scientists were forced to decipher the mind from the outside. And so they timed reflex arcs and imagined the link between the stimulus and the response. They talked about cognitive software and the irresponsible *id*, listened to patients discuss their dreams and primed undergrads with subliminal codes. This research program remains essential, of course; human nature, even after all these years, continues to astound and surprise.

But it's now possible to see *inside* that fleshy machine. In the last few decades, there has been a revolution in brain imaging. While Cajal was forced to stain dead tissue, we can now monitor the mind at work, using the flow of blood as a proxy for the activity of specific brain areas. Although brain imaging snapshots are imprecise, and their specificity is often misleading, they have profoundly changed our view of ourselves. For the first time, we can see the physical consequences of every thought, and watch, with more than a little awe, as the water of the brain becomes the wine of the mind.

There are no miracles here, though. While functional magnetic resonance imaging has seized the public imagination—those pretty pictures filled with splotches of primary color are hard to resist—modern neuroscience has been blessed with an astonishing array of new techniques for visualization and interpretation, from laser scanning microscopy to the supercomputer simulations of the Blue Brain. As a result, we are beginning to understand how thinking happens at its most fundamental level. Instead of mapping the flux of blood, we can create diagrams of the circuit itself, that vast electrical infrastructure designed by our genes and shaped by our experience.

The question, of course, is what to do with all this new information. What can we hope to learn from these latest scientific self-portraits? The only answer is that nobody knows. Perhaps we will learn everything. Perhaps the paradoxes of the mind will be solved by fluorescent neurons and high-throughput mRNA screens. Perhaps, in a few decades, even the hard problem of consciousness will seem like an easy puzzle.

Or maybe these images are only the beginning. Maybe they will slowly shift modern neuroscience away from its deductive model of research, in which researchers begin with a testable hypothesis, to a more inductive model, in which we first observe and stare and ponder. After all, the power of these pictures is that they allow us to observe the brain directly, without the frame of a conjecture. Such an approach might seem anachronistic—it was the model favored by Darwin and the Victorians—but when it comes to deciphering the mysteries of the brain it might be necessary. The human cortex is the most complex object in the universe: Before we can speculate about it, we need to see it, even if we don't always understand what we're looking at.

Finally, a note on the prettiness of these neural snapshots: It is not an accident. There is a grand tradition of scientists making art out of human anatomy, from the comic grotesqueries of Vesalius to the exquisite line drawings of Cajal. The twenty-first century is no exception. Just because these images depend on expensive machines doesn't mean the scientist has become a passive observer, or no longer thinks about aesthetics. Keats knew that truth exists in a tangled relationship with beauty, and nothing illustrates that poetic concept better than these scientific images. Their empirical power is entwined with their visual majesty. When we stare at these cellular images, we can't help but feel a twinge of recognition, the same twinge we feel when standing in front of a Rembrandt self-portrait. This, right here, is human nature, staring back at us. We have been laid bare.

PREFACE

There is no portrait of the mind. An army of curious men and women has endowed us with an abundance of infinitesimal glimpses into what is going on upstairs, yet truth be told I find the entire matter as baffling as the next fellow. Thankfully, we are not entirely in the dark. Prod the delicate matter in the head in the appropriate manner, and it just might reveal a small but important flash of insight, a clue among countless other clues. Prod by prod, glimpse by glimpse, we can begin to form theories about brain structure and function; thus, the history of neuroscience is the history of the techniques we employ to delve into the brain. Our entire edifice of knowledge, our very ability to pose questions about this organ and its relationship to the mind, depend on the tools and methods we have conceived to interact with them. Instead of delving into the fragmentary knowledge we have acquired concerning the workings of brain and mind, I propose that we discuss the techniques developed to study them and the gallery of portraits they have begotten over the centuries.

✳ ✳ ✳

Inside every scientific paper the reader will encounter an entire section, sometimes pages long, describing the methods that were employed to perform the experiments that are summarized in the article. Even by the standards of scientific communication—often noted for its dry, objective tone—the methods section is virtually platonic in its purity and remove. Sometimes I speculate that this stems from some unconscious desire to preserve the romantic notion of science as a wholly objective, unassailable, superhuman enterprise. Whatever the reason, the manipulations as described in the methods section are invariably performed by a nameless, faceless third-person-passive-form agent. One could be forgiven for wondering whether the procedures occurred overnight, ex nihilo, to the elation of a team of flesh-and-blood researchers who strolled into the lab the next morning only to discover the findings neatly summarized on their desks. Formally speaking, the methods section is somewhat of a downer.

At first blush, then, technique isn't exactly the stuff of exhilarating narrative. It offers no charismatic, eccentric, visionary, or charmingly but fatally flawed personality to hang a scientific story on. Knowing the ins and outs of experimental methodology won't make you a better person, or even sound clever at a cocktail party; it certainly won't help you know thyself with any more clarity. So why bother?

In early 2008, an improbable assembly of New York City writers and scientists, as well as myself, began asking this question. The setting was the living room of Stuart Firestein and Diana Reiss—researchers in neurobiology and animal cognition, respectively. The occasion was the inaugural meeting of an informal writing workshop we later called NeuWrite. Still going strong today, the group is focused on developing novel means of communicating science to nonscientists. One area that Stuart identified early on as sorely missing in most popular accounts of science—and thus fertile territory for new exploration—is the infamous methods section.

On Stuart's urging, our first writing exercises as a group was to take a raw methods section from a paper that we were particularly fond of for technical reasons (Buck & Axel, *Cell*, 1991, for the aficionados) and attempt to translate it into laymen's terms. It proved to be an exceedingly difficult task; looking back, I realize that we probably weren't terribly successful at it. As the workshop developed over the months we moved away from that first experiment and into many different directions. But Stuart's challenge remained in the back of my mind. When the opportunity arose that summer to create a book for Abrams that would focus on the art of raw neuroscience data, I realized that this medium would provide a perfect setting in which to revisit this challenge and indulge in a no-holds-barred foray into the techniques that produced the images in these pages.

Many of these pictures (and others much like them) are employed in the daily work of scientists around the world, yet are rarely made available, either as scientific evidence or as art, to a general audience. I hope that the reader will find, as I do, that they are intuitively beautiful and require no explanation for some degree of appreciation, but I also wish that their compelling aesthetic qualities will invite inquiry into what they show and especially into how they were obtained. For if the images are extraordinarily beautiful, I would argue that the principles underlying the techniques that created them are in some instances even more exquisite. The manifestations of beauty that we have been licensed to encounter in every nook and cranny of the brain depend, in every single case, on the cleverness and elegance of the strategies employed to illuminate them. My goal is for the reader's enjoyment of the images to be accompanied by equal appreciation for the craft that underlies their genesis.

As Western art developed over the past century, the concept behind the physical art object acquired a status on par with—or, some might argue, superseding—the thrilling of our senses by the object itself. Sol LeWitt's perspective—"the idea becomes a machine that makes the art"[1]—urges us to enjoy the aesthetic value of a concept independent of the actual item it engenders. By analogy, I often find the idea behind a technique to be irresistibly beautiful in and of itself. I propose then, that in science as in art we should delight not only in the physical manifestations of the data—examples of which fill these pages—but also in the ideas that produced them.

* * *

With very few exceptions, the most elegant—and fruitful—neuroscience methods share a common principle: faced with the task of sorting out the daunting tangle of neurons and their incessant chattering, successful techniques tend to selectively restrict the vision of the scientist, rather than record everything under the sun. If this sounds counterintuitive, consider the way that our visual system works. At any given moment we are aware of only a fraction of the details in our visual field: The brain excels at focusing on and extracting only the most meaningful signals flowing into the eye, providing us with only the information we need in order to react nimbly to events unfolding in the world around us. One hundred dots arranged in a line will always appear to you as a line, no matter how hard you try not to see it that way. The reason is that edges are a common feature of things you might need to deal with in the course of your lifetime. So your visual system is in the business of detecting edges, like the contours of a tiger stalking you in the jungle. What the brain doesn't do, however, is capture all of the information flowing in: The details of the moss growing on the tree trunks surrounding the tiger undoubtedly will go unnoticed as you evaluate the pros and cons of hiding or running away. For in this particular situation, moss data would not be of great use to you. Conveniently, your brain automatically filters it out, preventing confusion and helping you to solve the more pressing problem of saving your skin.

Less information is far more useful, so long as it is good information. In one way or another many of the most successful neuroscience techniques perform an analogous filtering. They enable the researcher to zero-in on very specific features in nervous tissue, cutting out most of the overwhelming mess and preventing it from distracting her from focusing on the question she seeks to answer. By imparting such a selective restriction of vision, the most ingenious methods have time and time again opened up formerly unsuspected universes.

✳ ✳ ✳

A few words about the format of this book. Every chapter opens with an essay by a senior expert on the main themes that will be discussed, providing the reader with a bit of background for the images and concepts that follow. I have written a caption to accompany each image, describing what it shows and how it was obtained; for those seeking more information I have in many cases included a reference to a scientific article, a full citation to which can be found at the back of the book in the picture credits section. I have adopted the convention of crediting only those people directly responsible for the images in this book. In some instances, these individuals are the inventors of the technique under consideration but the reader should not assume this to be the case for all images. I chose to avoid all but the coarsest chronology in my description of the techniques, and I certainly do not claim to present an exhaustive survey; this is neither a history nor a textbook. If you feel your eyes glaze over where the text gets complicated and qualified, fear not, just move on—you won't be missing much. My nonscientist friends tell me that the book is a bit of a workout to read. So it was to write, but I hope your and my delight will be worth our efforts.

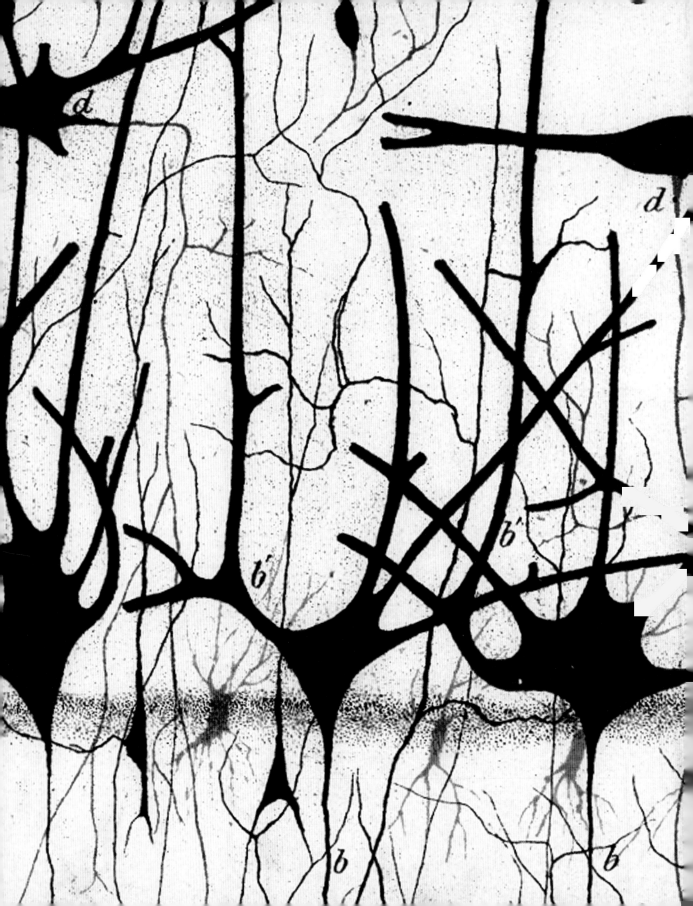

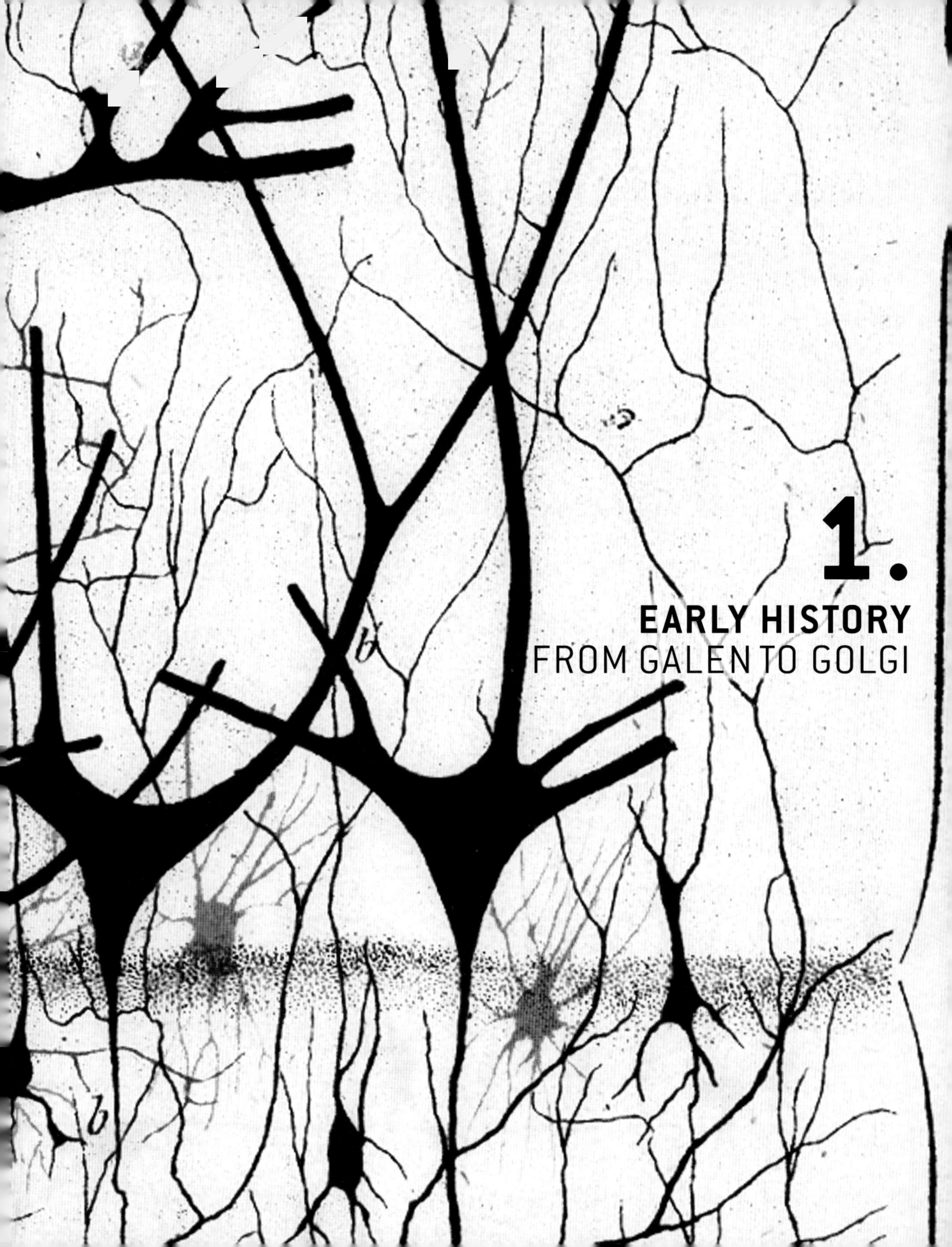

1.
EARLY HISTORY
FROM GALEN TO GOLGI

EARLY HISTORY: FROM GALEN TO GOLGI

by Nicholas Wade
University of Dundee

There are limits to what the naked eye can see. For centuries, students of the brain have struggled to overcome this frustrating restriction by developing ways of enhancing our vision, bringing the brain into sharper focus and thus enabling us to probe beneath its gray, amorphous exterior. A vast array of sophisticated methods and technologies has been invented, from microscopes to magnetic-resonance imagers (MRI scanners), not to mention a veritable armory of molecular and genetic devices. These have unveiled fascinating and previously unknown brain architectures and processes, and have granted us ever deeper access into the recesses of this dauntingly complex organ. But each new technique and each novel piece of equipment offers a view quite removed from the direct experience of our unaided eyes. The fact remains that the cells that make up the nervous system can only be seen with magnification (one step removed) and only when stained with special chemicals (two steps removed) that illuminate the imperceptible. This means that our perspective on the world of the brain is entirely dependent upon the nature of unseen, and in some cases, poorly understood, biochemical reactions and is mediated only by the technologies we have invented to view it.

The fruits of these expeditions into the invisible world were initially hand-drawn by scientists laboring over the lenses of early microscopes, trying to capture their observations with pen and ink. Now microscopic images are photographed instead of eked out by hand, but it is still striking how indirect, how far removed from the object itself, is our understanding of the structures of the brain. How does this indirect experience relate to the raw tissue of the brain? The delightful depictions of the cerebral organ in this chapter grapple with precisely this problem, and in a sense they also suggest an answer. The drawings here tend to represent two eyes, along with two hemispheres and two visual pathways and encourage one to seek an analogy with studies of binocular vision. Leonardo da Vinci, whose talent for drawing was matched by his prodigious skills at dissection, appreciated the difficulties of depiction centuries ago. He distinguished between natural perspective, what we perceive with the full use of our two eyes, and artificial perspective, in which angles are represented from a single, static viewpoint. Da Vinci summarized the problem when he wrote, "A Painting, though conducted with the greatest Art and finished to the last Perfection, both with regard to its Contours, its Lights, its Shadows and its Colours, can never show a *Relievo* equal to that of the Natural Objects."[1] In other words, the illusion of depth in a painting, no matter how masterfully rendered, will never match the actual perception of depth we experience in our daily lives. It is clear from his notebooks that he struggled long and hard with the contrast between monocular and binocular vision, and the problem of accurately representing reality with pencils, paints, and paper—tools that could never accurately reproduce the complexity of the real world.

The history of much of experimental science is one of gaining access to the unseen and of representing it in a medium—whether a sheet of paper or a computer screen—that inevitably fails to simulate our natural experience of the world. Consequently, this history is littered with mistakes, traps, and erroneous judgments regarding what is "real." Those scientists we now deem successful are the ones who were able, at times seemingly miraculously, to transcend the limitations of the methods and the media of their eras.

✳ ✳ ✳

The great anatomist Galen of Pergamum, who practiced medicine in second century Alexandria and Rome, performed groundbreaking dissection studies which served as the foundation for Western anatomical knowledge for more than a thousand years. Owing to severe restrictions on human dissection in the early Christian and Islamic worlds, Galen's accounts were mostly taken on faith and passed down dogmatically—mistakes and all. For centuries, studying the anatomy of the brain meant studying copies of his oft-misguided drawings. During this time, it became common for authors to take the liberty of rendering their theories of mind within putatively anatomical drawings, creating images that reflected not only the organ's shape but also the author's mistaken notions about how the mind functions, mingling physical form and psychological theory. This era in the history of brain science came finally to an end under the scalpel of the sixteenth-century Flemish anatomist, Andreas Vesalius.

Vesalius received Galen's anatomical ideas via the Islamic world, where scholars translated books from Ancient Greece, Rome, and other regions into Arabic, expanding upon and adding to the knowledge they found in these texts. In the fourteenth century, these translations, along with the original work of brilliant Islamic scholars such as Abu Ali al-Husain ibn Sina (also known as Avicenna), began trickling into Europe along expanded trade routes. In the wake of this new knowledge, sanctions prohibiting dissection of human bodies were gradually relaxed, and the understanding of human anatomy slowly became rooted in the study of actual cadavers. By the time Vesalius was performing his popular dissections in the 1500s, the act of cutting open human corpses had come to be seen as something of a necessary evil. Yet descriptions of the brain remained highly inaccurate, and a millennium of Galenic dogma blinded even the finest practitioners of the anatomical arts to some important physical features while persuading them of the existence of other purely fictional ones.

A famous example of this affliction can be found in the case of the *rete mirabile* ("wondrous net"), a prominent network of blood vessels located at the base of the brains in oxen and pigs. Galen, observing the rete in these animals, assigned to it a critical function in his theory of mind—that of distilling the base "vital spirits" flowing up from the heart into the "animal spirits" that were thought to govern brain function. Over the centuries, Galen's readers surmised that he had described a human *rete* and accordingly discussed it in their own accounts of human dissection. The fact that we wholly lack anything resembling this structure did not prevent its inclusion in medical textbooks as late as the seventeenth century. Even da Vinci, blinded by the persistence of dogma, failed to challenge the prevailing view when he had the opportunity of dissecting humans.

And so the renaissance of anatomy is generally understood as beginning with Vesalius's 1543 treatise on the structure of the human body, *De Humani Corporis Fabrica*. In this revolutionary book, Vesalius attempted to break free from Galen's legacy by offering his own observations of anatomical structures rather than relying on those of the revered second-century physician, presenting a synthesis of the most advanced knowledge and techniques in the science and art of his day. *Fabrica* is illustrated by a series of beautiful drawings, which not only provided images of unprecedented detail but also helped standardize the pairing of words and figures in anatomical texts. Yet even Vesalius's most detailed accounts of the brain remained patchy and imprecise, due in part to the fact that it tended to be the last part of the body to be examined. A typical dissection would take place over the course of several days, after which time the brain's soft, and by then decaying, tissues were unlikely to yield much insight into their delicate, hidden structures.

※ ※ ※

As we are now aware, drawing the form of the brain from direct observation can yield only minimal knowledge about its function; when high-quality microscopes arrived on the scene, it quickly became apparent that much of the action takes place at scales far lower than what we can perceive unaided. Tools of simple magnification have been available for many centuries, but specially constructed instruments for observing structures invisible to the naked eye did not became popular until the early seventeenth century. When Robert Hooke examined cork specimens under his microscope in 1665, the universe beneath his lens so resembled monastic bedchambers that he simply named its individual components "cells."

Soon after Hooke named the building blocks of the body, Antony van Leeuwenhoek improved upon Hooke's magnifying device and set out to answer a centuries-old question about whether nerves are hollow. A leading physiological theory popular from even before Galen's time—not to mention a basic tenet of medieval Christian philosophy—held that animal spirits coursed through the brain, passing between its pronounced cavities (called *ventricles*) along tubular, and presumably hollow, nerves. Armed with an exceptionally sharp knife and his experience with dyes from his work as a draper, Leeuwenhoek succeeded in demonstrating that nerves were, in fact, solid on the inside. In an ideal world, this would have laid the concept of animal spirits to rest, but despite ample evidence to the contrary, its vitality was prolonged because of the perceived lack of an adequate alternative.

The pernicious theory of animal spirits may also have persisted so long because little attention was paid to microscopic observations of the late 1700 and early 1800s due to the notoriously poor resolving power of lenses of that era. The world would have to wait until the 1820s, and the introduction of powerful achromatic lenses, to receive reports of crisp images at the scale of individual cells. After that, the pace of discovery accelerated significantly. By the end of the next decade scientists had determined that all organic matter is composed of cells, and by the close of the nineteenth century, they had established that the entire nervous system is constituted of independent nerve cells.

Late nineteenth-century anatomists moved beyond simply cataloging the structures they encountered to formulating strong—and occasionally correct—theories about their functions. Unlike their medieval forebears, they could now base their theories on firm anatomical foundations rather than religious or philosophical ones. It was the dawning of modern neuroscience. But while the achromatic microscope ushered in a new perspective on the biological world, the untreated brain remains an undifferentiated gray slate no matter how powerful the lens. Nerve cells, when examined under a microscope, are difficult to see because their exquisitely fine parts barely alter a beam of light that passes through them. A chemical alteration of some sort is required to reveal these structures, so it is the combination of chemical stains *and* the means to examine them microscopically that truly fueled the modern study of the brain.

We return to the problem of having to modify the very chemistry of the objects we seek to study and to rely upon the faithfulness of our optics. Da Vinci's qualms about the reliability of indirect perception here are multiplied almost beyond recognition. For the reader, a further layer of mediation is added by the translation of the view from the microscope to the figure on the page. Paradoxically, it is only through this imperfect series of transformations that scientists and laypeople alike may gain access to the structures of the brain and begin to contemplate the mysteries of their function.

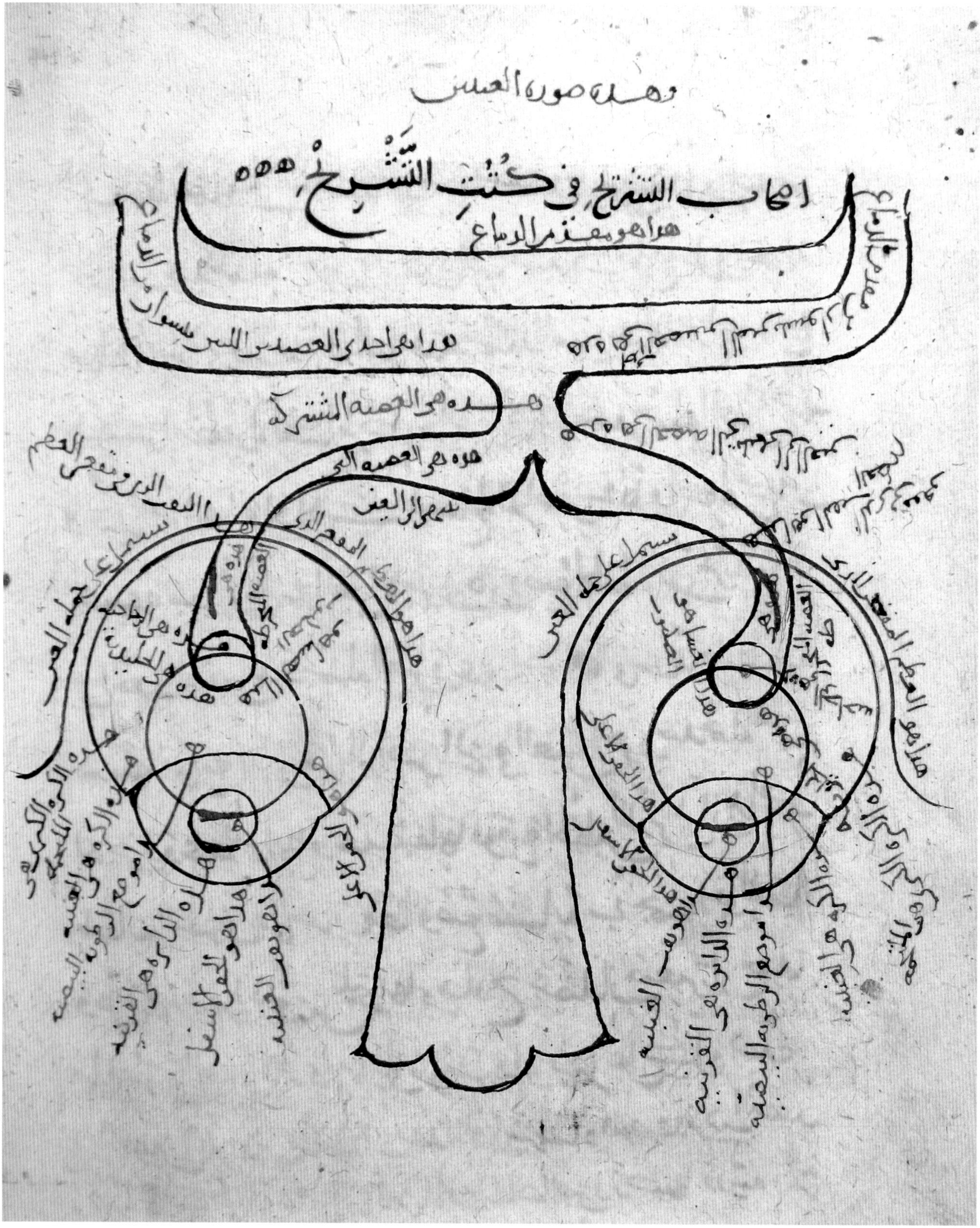
وهذه صورة العين
اصحاب التشريح في كتب التشريح
هذه هي العصبة المشتركة

Visual system.
Ibn al-Haytham, circa 1027
(published in 1083).

The oldest known depiction of the nervous system seems reassuringly well ordered: a large nose at the bottom, eyes on either side, and, flowing out of each eye, a hollow optic nerve that meets the other for a moment before parting ways and continuing on to the brain— all of this meticulously named and cataloged. From this unadorned sketch, drawn in eleventh-century Cairo by Ibn al-Haytham, comes a premise that is so elementary as to seem almost trivial: In the nervous system, information travels. From here we go on to formulate the more exciting—and more explicit—hypothesis: At each stop in its path, signals from the world outside are somehow processed, interpreted, or put to use. How information travels from one part to another inside the brain, and how it is processed at each step, is the business of neuroscience.

Al-Haytham's drawing was based in part on the teachings of Galen of Pergamum, a second-century Roman physician and anatomist who, although dead for eight centuries, continued to exert an outsize influence on the world of brain science. In ancient Rome, Galen worked in an environment of relative permissiveness when it came to cutting through bone and flesh to study the organs underneath, an activity that religious and legal restrictions subsequently rendered taboo for centuries thereafter. Since human dissection was off-limits, he employed as his subjects a veritable zoo—apes, dogs, bears, stags, camels, and one elephant. Otherwise, his exposure to human anatomy was restricted to wounded gladiators and rotting or deteriorated corpses that he came upon by accident.

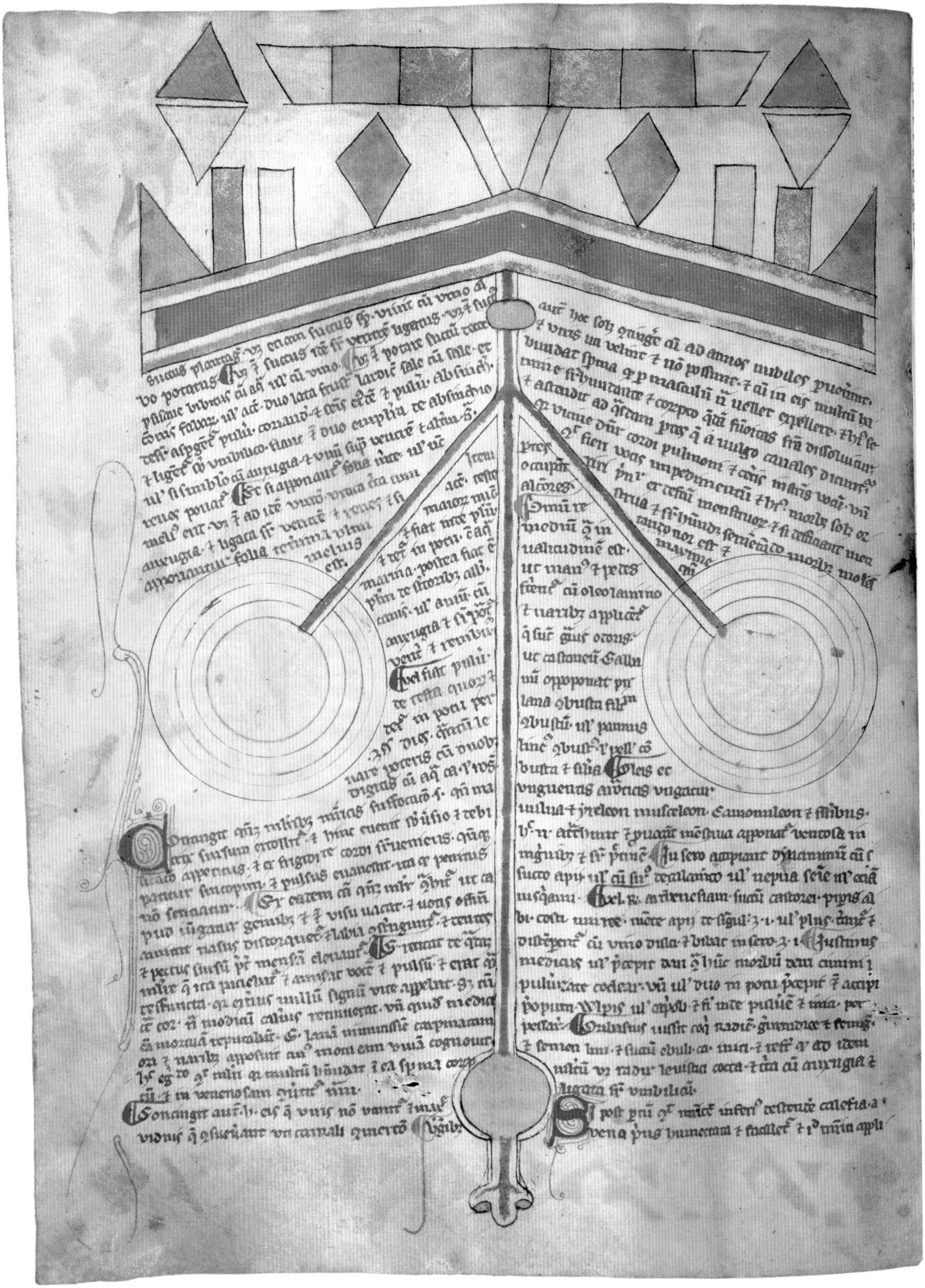

Ventricular theory.
Anonymous, England, 1292.

With religious authorities issuing blanket bans on dissection after Galen's era, many anatomists relied on interpretation of and commentary on his findings, which he meticulously cataloged in a series of hundreds of treatises. Many of these were lost in a fire, and if he did draw figures summarizing his findings, none are known to exist today. Thus, our vision of Galen is one that is necessarily mediated by the eyes and minds of his followers. His fragmented output, from an era more open to exploration, crystallized, amplified, and took on baroque characteristics of its own as it bounced around the libraries of Europe and the Middle East throughout the Dark and Middle Ages. His words were dissected under the pen; his theories were bent to accommodate the leading views of the time. Occasionally his views were challenged—especially in the Islamic world—but by and large his theories became dogma.

Shown opposite is an anonymous late-thirteenth-century depiction of many of the same structures in al-Haytham's drawing, although the nose is decidedly smaller. This drawing follows a view laid down by fourth-century church fathers who sought to integrate Galen's anatomical findings into the teachings of the church. They, like Galen, believed that the mind was the product of rarefied humors, or "animal spirits." But in a departure from Galen's belief that brain substance as a whole underwrote the workings of the mind, they localized its function to a series of cavities deep inside the brain called *ventricles*. (Their representation as rhomboids is fairly abstract in this example; the two drawings on pages 23 and 25 depict them more explicitly.) The church fathers' reason for focusing on the ventricles is a bit obscure; perhaps they interpreted their central location—comfortably ensconced within the soft protective tissue of the brain—as signifying their importance. Today, we know that these structures serve a less exalted purpose: They store the salty solution in which the nervous system bathes.

Woman's headdress.

Anonymous, Saxony, 1441.

This anonymous fifteenth-century drawing splices together elements of Galen, the church fathers, and a predecessor to both: the Greek philosopher Aristotle, whose output includes records of his avid study of nonhuman anatomy. Despite his reputation as a logician and a metaphysician, Aristotle's approach to investigating the universe was empirical: He trusted only those things he could perceive or study directly. (This is a lesson Galen took to heart and sought to pass along to his disciples, who, while stubbornly faithful to his anatomical treatises, seem to have ignored his methodological admonitions as the centuries rolled on.) Aristotle believed that the brain served merely as a cooling unit siphoning off the excess heat from the heart, where the rational soul did its work. In this drawing his theory is represented by the inclusion of the heart in the mind's functions, playing a part in the sense of touch and taste, alongside a Galenic (and church-father-ish) view of brain and ventricle function, as symbolized by the inscription of faculties— sensation, imagination, and memory—in the subject's headdress. The band around the woman's neck informs us that "touch is located in all parts of the body."

Although this particular drawing would seem to indicate otherwise, Galen had, more than a millennium before, punched a significant hole in Aristotle's cooling-unit theory when he set out to test one of its important tenets: The brain, in order to cool the heart, must itself be cold to the touch. How momentous must the day have been when Galen first cracked open the skull of an animal and rested his own hand on its living, pulsating brain—it was *warm*!— thus relegating the heart to baser functions and establishing the contents of the head as the favored subject in the study of the mind. Moreover, Galen determined that this brain received the nerves from sensory organs and, when it was damaged, could cause the loss of sensory and mental faculties. The evidence seemed overwhelming: The mind was all in the head, not the heart. Neuroscience begins with an answer to the question of *where*. Those were the days when a sharp knife and a vigorous wash of the hands were enough to establish theories with millennial shelf lives.

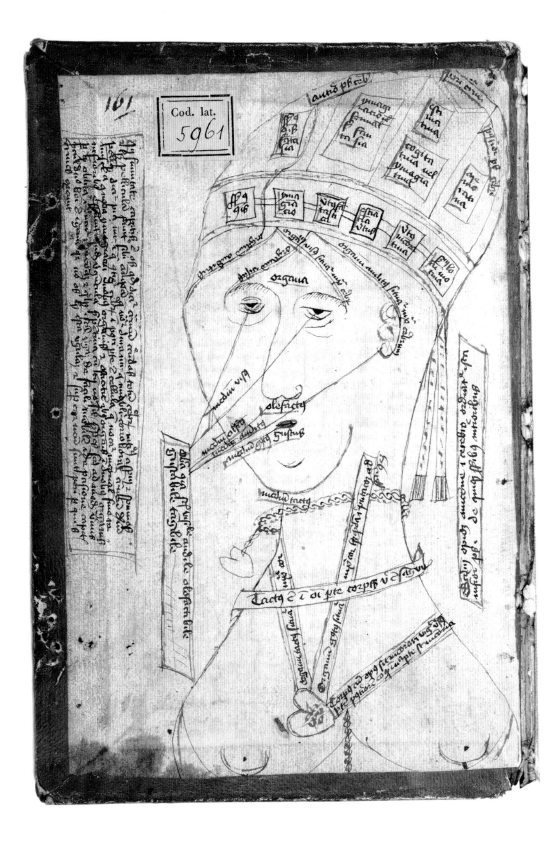

Mind and universe.
Robert Fludd, circa 1621.

This seventeenth-century illustration by the English physician Robert Fludd takes on a more metaphysical bent, relating the soul's faculties to realms such as the sensible, the imaginary, and the divine; a writhing *vermis*—Latin for "worm"—connects imagination (which overlaps with perception) and cognition (itself linked to judgment). This illustration's publication date is a testament to Galen's staying power: Although Galen's ideas were by now under fire from many parts, Fludd's figure suggests their strong influence even then, a full fifteen hundred years after their formulation. The blame for this staggering failure of brain science to advance in more than a thousand years after his passing lies more with the subsequent generations of men who were never permitted—or didn't bother—to practice anatomy with their own hands. They passed along many of his theories without examination, packaging his prescient insights with his most egregious errors.

One particular mistake of Galen's warrants mention, if only to serve as a cautionary tale for the program of empiricism that would soon be reintroduced into Europe during the Renaissance. When Galen investigated the point at which the two optic nerves—each one emanating from an eye—come together on their way to the brain, he noted that they *appeared* to cross over to the other side in some of the animal specimens that he had dissected. (The al-Haytham illustration on page 18 shows the chiasm clearly in the center of the diagram.) Yet despite the evidence before him, Galen rejected the notion that the optic nerves might cross, and later chalked up this impression to his own poor dissection rather than the actual state of things. The modern neuroscientist, aware that large portions of the optic nerve *do* cross at the chiasm, is reminded of the difficulty inherent in the interpretation of data. It is not always obvious when an observation is worth recording or simply the result of a flawed preparation; the secrets of the brain will never reveal themselves unambiguously.

PORTRAITS OF THE MIND

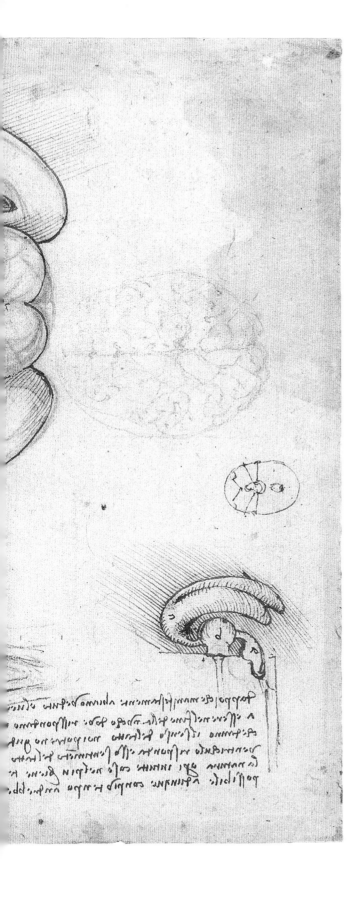

The shape of ventricles.
Leonardo da Vinci, circa 1508.

An early proponent of what would soon become a forceful return to the study of actual anatomical samples, the Italian polymath Leonardo da Vinci often conducted his dissections in secret, as restrictions began to be cautiously relaxed. Dissatisfied with Galen's description of the ventricles, and seeking a way to more accurately determine their structure, Leonardo realized that the brain was sufficiently complex that cutting and looking alone would not suffice. That the organ of interest was necessarily found in a cadaver—often one that had been lying around for a while—complicated things further, making brain study an eminently messy affair. To circumvent this problem and study ventricles in their unadulterated state, he turned to common statue-casting techniques, injecting melted wax into the ventricle of a freshly killed ox while siphoning off the excess fluid through holes poked in the other end. After the wax had cooled and solidified, he carved the surrounding brain away until nothing was left but the shape of the cavity. The top two diagrams here show his rendering of the cavity from both the vertical and horizontal view. If the history of neuroscience is a history of seeing, then Leonardo ushered in the conviction that our eyes alone cannot do the trick.

A renaissance in anatomy.
Charles Estienne, 1545.

Leonardo da Vinci would soon be followed by
a pair of terrifically skilled anatomists whose
singular output would mark the beginning of
modern anatomy: the Brussels-born Andreas
Vesalius and the Parisian Charles Estienne.
Owing to a sordid legal dispute with an associate,
the publication of Estienne's discoveries was
significantly delayed, and Vesalius, who
consequently scooped him by two years,
received some credit that perhaps should have
been granted to both. Thus, today, Vesalius
is the name we most closely associate with
the renaissance of anatomy. However, it is
Estienne's illustrations that give the viewer
pause: Notwithstanding their gaping flesh, the
anatomical subjects were often set in distinctly
nonclinical surroundings. Features that might
not find their way into today's anatomy
textbooks—melancholic facial expressions,
stylish haircuts—were outlined with precision
and grace. Perhaps this was a result of the
practice in their day of recycling images from
nonscientific publications to cut down on
production costs, or perhaps it reveals the
humanism with which both approached
their subject.

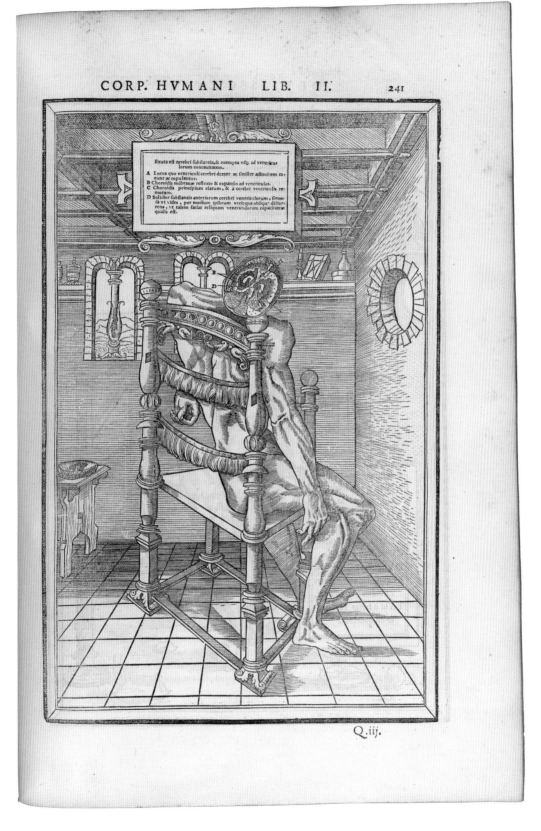

ANDREAE VESALII

BRVXELLENSIS, SCHOLAE
medicorum Patauinæ professoris, de
Humani corporis fabrica
Libri septem.

CVM CAESAREAE
Maiest. Galliarum Regis, ac Senatus Veneti gra
tia & priuilegio, ut in diplomatis eorundem continetur.

Frustrated by the limited opportunities during
his time in medical school to, as he delicately
phrased it, "put my own hand into the business,"[2]
Vesalius overcame any moral qualms—and the
occasional pack of wild dogs—to scour local
cemeteries for human remains. In response to
this bold departure from traditional anatomical
study, Vesalius's former mentor, Jacobus
Silvius, once implored his colleagues to "pay no
attention to a certain ridiculous madman, one
utterly lacking in talent who curses and inveighs
impiously against his teachers"[3] and urged the
powers that be "to punish severely . . . to
suppress him so that he may not poison the
rest of Europe with his pestilent breath."[4] Clearly,
Vesalius was onto something. Yet in a town
still in the thrall of Galenic theories, Vesalius's
unsettling obsession with performing his own
dissections raised eyebrows and led him to
settle in Padua, Italy, where his research
program received a greater measure of support.
There, he was supplied with a reliable stream
of corpses fresh off the gallows, whose deaths
were occasionally timed to accommodate his
professional needs. Vesalius's dissections
occasionally became massive public affairs,
bordering on the ceremonial, for the benefit of
students, colleagues, dignitaries, and assorted
gawkers. The frontispiece of his 1543 magnum
opus, *De Humani Corpus Fabrica (On the
Workings of the Human Body)*, seen here, hints
at the flavor of these investigations.

     It was his dissections in Padua, and
an epiphany induced by a side-by-side display
of human and ape skeletons, that put the
proverbial nail in the coffin on Galen's anatomy.
Vesalius realized that while his Roman predecessor had claimed to be characterizing
human anatomy, the structures he had
described more closely matched the anatomy
of certain animals—a fact that is hardly
surprising in retrospect, since in Galen's Rome
human dissection was prohibited. What Galen
had failed to account for was the possibility
that his findings in apes, stags, and dogs might
not prove true for humans, as well.

The year 1543 was a watershed in the European
science publishing business, with the simultaneous release of Vesalius's *Fabrica* and Nicolaus
Copernicus's *On the Revolutions of Heavenly
Bodies*. In the space of a few months, the stage
was set for the modern study of two universes,
those within and without. Vesalius's volume
was a massive, lavish tome containing exquisite
plates believed—although never proven—to have
been drafted by a student from Titian's studio.
Among other achievements, Vesalius provided
a precise account of the shape and position of
ventricles in humans (overleaf left), and although
he stopped short of overturning the theory
that animal spirits were located therein, he did
express puzzlement over the similar appearance
and size that they took in nonhumans. Despite
this committed study of ventricular properties,
it is clear that he did not overly concern himself
with the shape of brain substance itself; closer
in appearance to an intestinal tract, the brain
shown here (overleaf right) bears little resemblance to the actual folds familiar to any first-
year medical student today. The draftsmen,
who did not detect these characteristic patterns,
drew what they saw: a big gray mess encased
in the skull. Until the seventeenth century, in
fact, these folds were assumed by many to be
involved mainly in the secretion of phlegm and
tears; all the action was still believed to take place
in the ventricles. Though Galen's anatomy had
by now been reduced to tatters, his ideas still
cast a long shadow.

QVARTAE FIGVRAE, EIVSDEMQVE CHA-
racterum Index.

IN quarta figura omnes duræ tenuisq; membranaru partes resecuimus, quæ in prioribus figuris occurrerunt, ac dein dextrā sinistramq; cerebri portionē sectionis serie ita ademimus, ut iam cerebri uentriculi in conspectū uenire incipiat. Primum nanq; secundùm dextrū callosi corporis latus, ubi sinus altero *M* in tertia figura notatus cōsistit, longā duximus sectionē, quæ per dextrū cerebri uentriculū ducta, dextræ cerebri partis eam portionē abstulit, quæ supra sectio nē habebatur, qua orbiculatim caluariā serra diuisimus. Atq; quum idem quoque in sinistro la tere absoluimus, hic ad sinistrū latus sinistrā cerebri partē ita reposuimus, ut superiorē sinistri uentriculi sedem aliqua ex parte cōmonstraret, calloso interim corpore in capite adhuc seruato.

A, A, *A* Cerebri adhuc in caluaria relicti pars dextra. B, B, B Pars sinistra.

C, C, *C* Portio cerebri sinistra, quæ sectionis serie à reliquo cerebro ablata, hic resupina iacet.

D, D, *D* Lineæ partim cerebri anfractus, et partim uariū cerebri substātiæ colorē cōmonstrantes. quicquid enim extra lineas consistit, quasi luteū & subcinericiū magis est: quicquid intra, ad a

E, F. mussim album uisitur. quemadmodum *E* & *F* in dextra & sinistra cerebri parte luteum est: *G*

G, H. uerò & *H* album prorsus, & interdum rubris punctis interstinctum.

I, I Callosum corpus utrinq; à cerebri substantia, cui alioqui continuatur, liberum.

K, K Callosi corporis portiuncula, adhuc sinistræ cerebri parti quæ adempta est, continua.

L, L Dexter cerebri uentriculus. M, M Sinister cerebri uentriculus.

N, N Sinistri uentriculi superioris sedis portio.

O, O Plexus cerebri ab imagine quā cum extimo fœtus inuolucro similem exigit, χοροαδής nuncupatus.

P, P Aranearum modo graciles uenæ dextri & sinistri uentriculi substātiæ hoc in loco connatæ, ac ab illis diductæ uasis, quæ nuper commemoratū, & secundis nō absimilem plexum extruunt.

Q Venulæ à nuper quoq; cōmemoratis uasis in tenuē cerebri membranā sub callosi corporis ante riori sede huc excurrentes, & incerta serie (quemadmodum & *P* notatæ) inter resecandum se offerentes.

QVINTA SEPTIMI LIBRI FIGVRA.

PRÆSENS figura quòd ad relictam in caluaria cerebri portionē attinet, nul la ex parte uariat: atq; id solū habet proprium, quod callo sum corpus hic anteriori sua se de à cerebro primùm liberaui mus, ac dein eleuatum in poste riora refleximus, septum dex tri ac sinistri uentriculorum di uellentes, & corporis instar testudinis extructi superiorem superficiem ob oculos ponētes.

Ab A *A, A, A* itaq; & *B, B,*
ad Q. *B,* ac dein *D, D, D,* & *E* & *F,* & *G* & *H* eadem hic indicant, quæ in quarta fi gura. Sic quoque & *L, L,* & *M, M,* & *O* & *P* & *Q* eadem insinuant.

R, R, *R* Notatur inferior callosi corporis superficies. est enim id à sua sede motum, atque in poste riora reflexum.

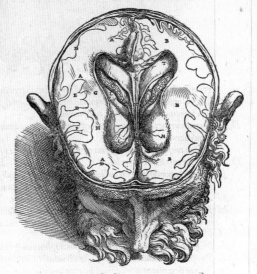

S, T, V Supe

parte adapertus, uerum elatiorem ipsius costam uti naturaliter se habet integram, & quartæ
circuli partis modo extuberantem ostendens.

D, D,   D   Duo inuicem appositi ductus, uenarū modo in duram cerebri membranam secundùm ipsius
uniuersum latus excurrentes.

E   Ductus duræ cerebri membranæ, in quem sexta caluariā ingrediens uena exhauritur.

F, F, F   His characteribus uenulæ indicatur, à dura cerebri membrana per caluariæ foraminula ad capitis cutem, membranasq́ caluariam succingentes, transmissæ, quarū frequentiores & crassio-
res iuxta F maximè latitantis sedem plerunq́ obseruantur.

G, G, G   Portiunculæ fibrarum à dura membrana per coronalem suturam ad caluariā succingentis mem
branæ constituttonem prosilientium.

H, H   Portiunculæ fibrarum quibus sagittalis sutura uiam offert.

I, I   Hi quoq́ characteres in umbra occipitij regionis latitāt, sedem notātes, à qua fibræ porrigun
tur per suturam Λ Græcorū similem, ad caluariæ inuolucri constitutionē deductæ.

K   Vnum tuberculū eorū quæ inæqualibus caluariæ sinibus plerunq́ iuxta sagittalis suturæ cū co
ronali coitū cōspicuis, adnasci solēt. Caput ex quo hanc primā depinximus figurā, tribus eius ge
neris donabatur tuberibus, quorū unum K insigniuimus, & utrinq́ ad H unū quoq́ se offert.

L   Cauitas frontis ossi iuxta superciliorum sedem propria, quæ inter secandum subinde aperitur, si
quando frontis os non procul à superciliys serra diuiditur.

SECVNDA SEPTIMI LIBRI FIGVRA.

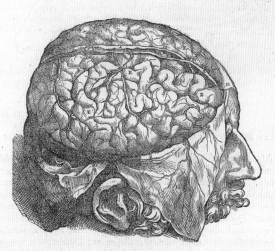

SECVNDAE FIGVRAE, EIVSDEMQVE CHARA-
cterum Index.

PRÆSENS figura sectionis serie primam subsequens, tertium duræ membranæ
sinum (quem prima figura C aliquot insignitum gerit) longa sectione secundùm capitis longitu
dinem ducta adapertum commonstrat. Insuper ad huius tertij sinus latera, per capitis quoq́ lon
gitudinem duas deduxi sectiones, utrinque nimirū ad sinum singulas, quæ durām membranam dun
taxat penetrarunt, & duræ membranæ latera ab ea membrana separarūt parte, quæ dextram
cerebri partem à sinistra dirimit, atque in subsequēti figura tribus D insignietur. Præter tres
iam cōmemoratas sectiones utrinque aliā quoque molitus sum, quæ ab aure ad uerticē pertingēs,
solam

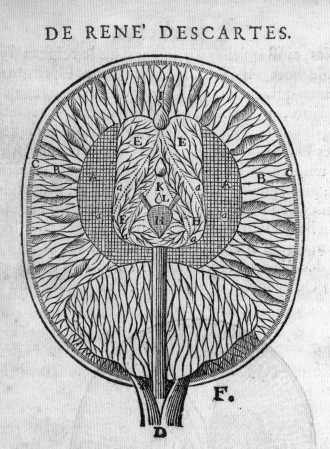

Notez auſſi que lors que ie dis que les Eſprits en ſor-
tant de la glande H , tendent vers les endroits de la ſu-
perficie interieure du cerueau , qui leur ſont le plus di-
rectement oppoſez , ie n'entens pas qu'ils tendent toû-
jours vers ceux qui ſont vis à vis d'eux en ligne droite;
mais ſeulement vers ceux , où la diſpoſition qui eſt pour
lors dans le cerueau les fait tendre.

Or la ſubſtance du cerueau eſtant molle & pliante,
ſes concauitez ſeroient fort étroites , & preſque toutes LXV.
Quelle dif-
feren:e il y

I ij

Pineal gland.
René Descartes, 1664.

In the mid-seventeenth century the philosopher and mathematician René Descartes introduced a fresh set of ideas when he set out to examine the mind-body problem: How can the intangible mind causally interact with the physical body, receive sensory impressions, and govern its motions in turn? This paradox over how the immaterial mind could relate to the physical world was particularly difficult for a thinker who had set out to understand the universe through the prism of physical mechanics. Inspired in part by lifelike statues that were animated by sophisticated hydraulic systems in the royal gardens of the Saint-Germain palace on the outskirts of Paris, he sought to understand all phenomena, including animal and human behavior, according to mechanistic explanations. To reconcile this worldview with the impalpable quality of the mind, Descartes felt compelled to identify the anatomical interface between the two, and he settled on the *pineal gland* (labeled "H" in this drawing). He posited that this gland shuttled information from the physical world up through the nerves and into the immaterial mind; the gland could also, in reverse, transmit the mind's commands back down to the body. For Descartes and his contemporaries, who were inclined toward theories of mind involving animal spirits in the ventricles, the selection of the pineal gland as the conduit between mind and matter presented the obvious advantage of being located directly beside the ventricles. (Descartes's prodigious skills as a philosopher, however, did not cross over to dissection, and he produced a notoriously erroneous anatomy; his ideas regarding the exact location of the pineal were quickly disproven.) Descartes's pineal hypothesis was further supported by his belief that it was the only single structure in an organ where all others are mirrored on two sides: perfect for interacting with the unitary mind. Although his purported solution to the mind-body problem now seems wanting—and was assailed immediately upon delivery—his greater vision of the mechanical nature of the body and its nervous system resonates to this day. Where he made analogies to clever hydraulics, we now appeal to computer chips, but the broader message remains: The nervous system obeys physical laws—laws that, if properly understood, will dissolve the mist and yield insight into the system's function.

View of a human brain from below.
Thomas Willis and Christopher Wren, 1664.

Descartes's vision of the nervous system as mechanism would find its experimental realization across the English Channel in the work of Thomas Willis and Christopher Wren. Willis, who wielded the scalpel with consummate virtuosity, generated a mass of anatomical knowledge that set the new standard for neuroanatomy after Vesalius. Wren is perhaps best known for having designed Saint Paul's Cathedral in London, but when he wasn't drawing up building plans, he found the time to make significant contributions to the fields of mathematics, astronomy, and—assisting Willis—anatomy. Together, they synthesized their observations into a single, coherent, three-dimensional whole, as shown here. With human dissection now increasingly mainstream in most of Western Europe, they provided an account of neuroanatomy that was orders of magnitude more precise than Vesalius's, shifting the emphasis away from the ventricles and back to brain substance itself.

The ventricular theory fell apart when Willis, who pioneered novel dissection methods noticed that the ventricles collapsed the instant the organ was removed from its case. Concluding that the ventricles were merely "a complication of the brain infoldings,"[5] Willis turned his interest to the arterial system, the dense web of blood vessels that innervates the tissue. He believed these served as the conduits for a "nervous juyce" of animal spirits sloshing around in the brain. In order to examine the vasculature's labyrinthine structure, Willis would inject a dyed solution into an artery and follow its tortuous path through the brain, echoing Leonardo's prescient method for studying the ventricles, and confirming once again that the unaided eye would not be sufficient to the task of studying this organ. His selective method of observation, which illuminates arteries and relegates the rest to the background, finds a direct expression today in *cerebral angiography*, a clinical method discussed in Chapter 7.

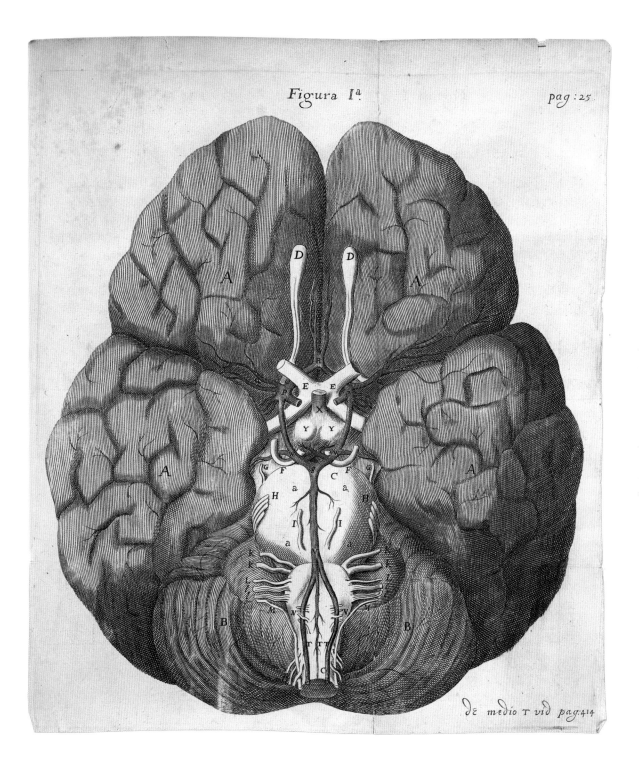

Human skull inscribed by a phrenologist.
Anonymous, nineteenth century.

It is worth pausing for a moment to recognize that the first question of neuroscience up until the seventeenth century had been about location—*where* the mind was to be found in the human body. The query would be bizarre if it weren't so commonplace. This approach makes intuitive sense when discussing other organs: the heart pumps blood; this part performs one function; that part, another. But why must the default assumption concerning something as elusive as the mind be that it can be assigned to a physical, clearly delineated set of structures?

Willis, in a stage-setting move, divided the brain into discrete parts: the inner (and lower) *brain stem* for basic functions such as breathing and heartbeat; and the surrounding outer "shell," or *cerebrum*, for more exotic ones like memory and volition. He did not go so far as to assign specific faculties to distinct areas of the cerebrum, but that notion was bound to emerge, and sure enough it did, but not until the mid-eighteenth century. Its initial author, the mystic Emanuel Swedenborg, was sufficiently isolated from the scientific mainstream that his theory did not take hold until decades after it was formulated. It found its first viable champion, at the end of that century, in Franz Joseph Gall, who claimed to have had inklings about his theory from the tender age of nine.

Termed *phrenology*, Gall's logic seemed flawless: (1) Our mind's moral and cognitive traits are biologically derived and governed by the cerebrum, which (2) is itself compartmentalized into suborgans that specialize in a variety of mind-related business such as memory, pride, a tendency to murder, poetic talent, and so on. These parts grow according to the use that is being made of them, thus (3) causing the surrounding skull to adapt its shape around the bumps and crevices of the brain beneath. It follows, in conclusion, that the contours of the human skull can provide a direct readout of the hidden intellectual, moral, and emotional tendencies of the mind inside; violent criminals and celebrated geniuses were said to have telling skull traits that betrayed their mental and moral faculties. This photograph of a human skull, inscribed by a phrenologist, shows areas associated with different parts of the mind. It is easy today to chalk up Gall's reasoning to quackery, and its service to subsequent commercial and ideological uses (such as racism) certainly does not help his case. But Gall ushered in the first modern theory ascribing different mental functions to different parts of the cerebrum. While he was entirely off the mark on the details, his paradigm guides us to this day—only the coordinate system has shifted to indicate positions within the brain rather than on the skull.

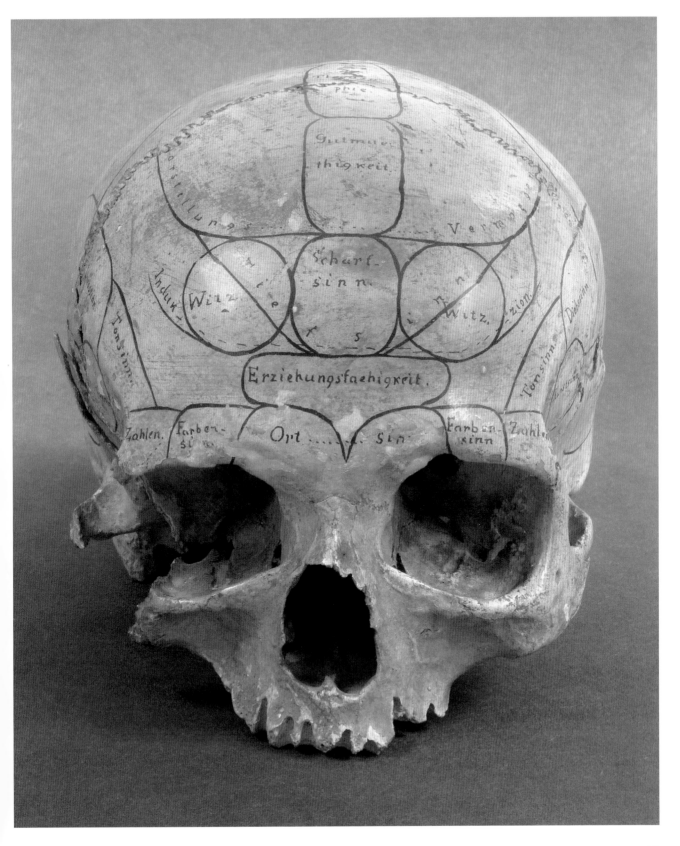

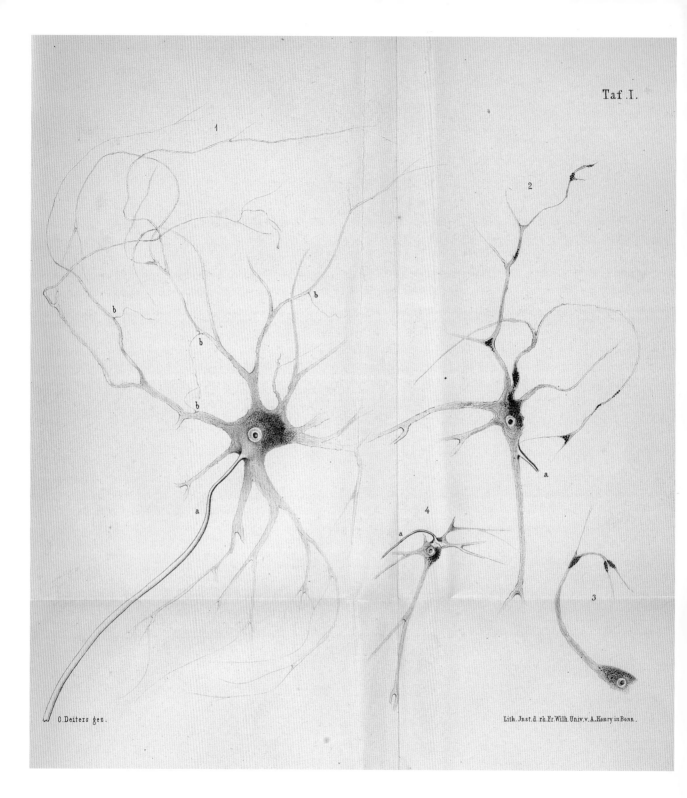

O.Deiters gez.

Lith. Jnst.d.rh.Fr.Wilh.Univ.v.A.Henry in Bonn.

Anatomy of the neuron.
Otto Friedrich Karl Deiters, 1865.

Neural explorers charting out brain territory would soon discover that the substance they studied, the very ground they stood on, was actually an amalgamation of infinitesimally small parts, or *neurons*. The mid-nineteenth century witnessed the arrival of powerful new microscopes that could reveal anatomical structures so small they had been previously indistinct, if visible at all. In many non-neuronal biological tissues, some cellular structures stand out unaided. But the brain, ever resistant to study, appears only as a pale blank slate. Two options remain for isolating and studying the structures of this tissue: Either they must be painstakingly dissected, or they must be made to stand out through chemical staining. The first strategy enjoyed a short period of attention but quickly fell out of favor, as chemical procedures were developed to do the seeing for us. It required staggering surgical genius to manually tease apart beneath a microscope neurons whose parts are a mere fraction of the size of a human hair. That feat alone merits recognition before we turn to those methods that are accessible to the rest of us: With nothing but fine instruments, sharp eyes, and a dazzling hand, the German Otto Friedrich Karl Deiters picked apart the pieces and became the first to describe, in the mid-nineteenth century, the principle components of the neuron. Seen here in his exquisite drawing is a cell body in the center (the *soma*), a thin *axon* running out one end, and a web of *dendrites* protruding from the other.

Plate 3

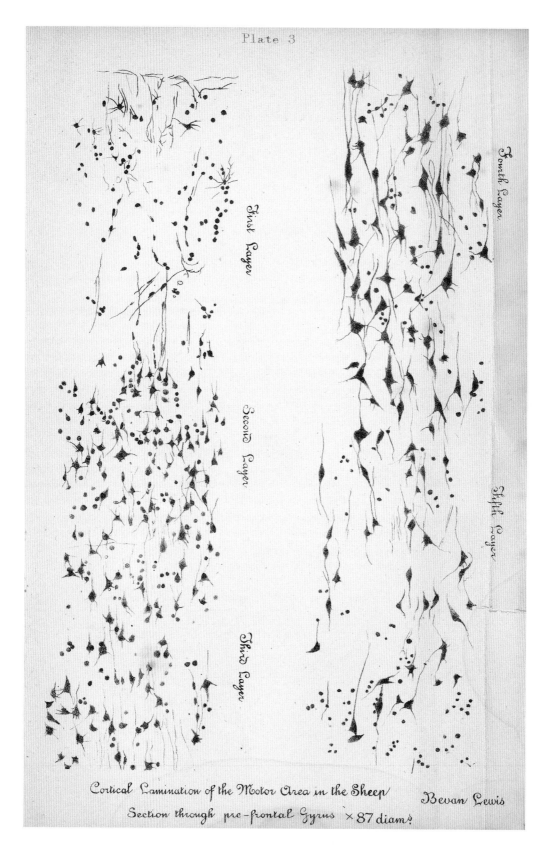

First Layer

Second Layer

Third Layer

Fourth Layer

Fifth Layer

Cortical Lamination of the Motor Area in the Sheep

Section through pre-frontal Gyrus ×87 diams

Bevan Lewis

Neuron layers in the neocortex.
William Bevan-Lewis, 1878.

The history of chemical staining goes back at least to the seventeenth century, when the Dutch microscopist Anton van Leeuwenhoek splashed a mixture of saffron and "burnt wine" (or brandy) onto his muscle specimens. But the second half of the nineteenth century proved extraordinarily fertile, as scientists met the challenge of microscopy with a plethora of novel staining methods for biological tissue. As patterns in previously indistinguishable nervous tissue came into focus, a second set of neural cartographies emerged to complement the level of organ and ventricle. The legacies of da Vinci and Willis and Wren were at last set into full motion, now on a physically smaller scale: Success in the study of the brain would henceforth depend on clever ways of staining or isolating only those structures of interest, allowing the rest to fade into the background.

But interpreting the patterns revealed by chemical stains is by no means a trivial exercise, for once the object of study has been chemically altered, it is not always easy to distinguish the real from the artifact. The next three drawings are examples of the data obtained from observing chemically stained samples under the microscope. They are, by nature, interpretations and simplifications of what appears in the microscope. Yet even as such, they suggest the magnitude of the challenge that lay ahead: Within each functionally defined area of brain, there exists an intricate network of centers and roads. A century later, we find ourselves a little less lost, but our cartographies of brain areas, and the patterns made up by their cells, remain hidden in many places. This drawing (opposite) by the British physician William Bevan-Lewis— who was interested in the link between brain cells and mental disease—is from an 1878 paper describing the differences between the cells that make up the now canonical layered aspect of the neocortex.

Hippocampus.
Joseph Jules Dejerine, 1895.

This is an early cartography of the *hippocampus*, an area of the brain critical for learning and memory, as interpreted by the French neurologist Joseph Jules Dejerine.

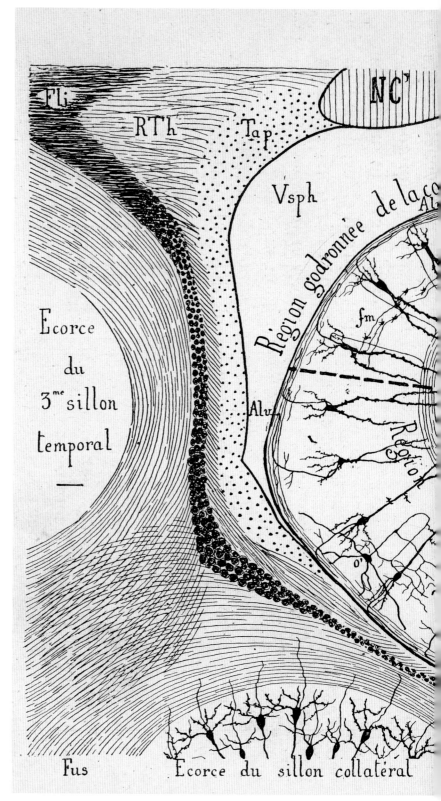

PORTRAITS OF THE MIND

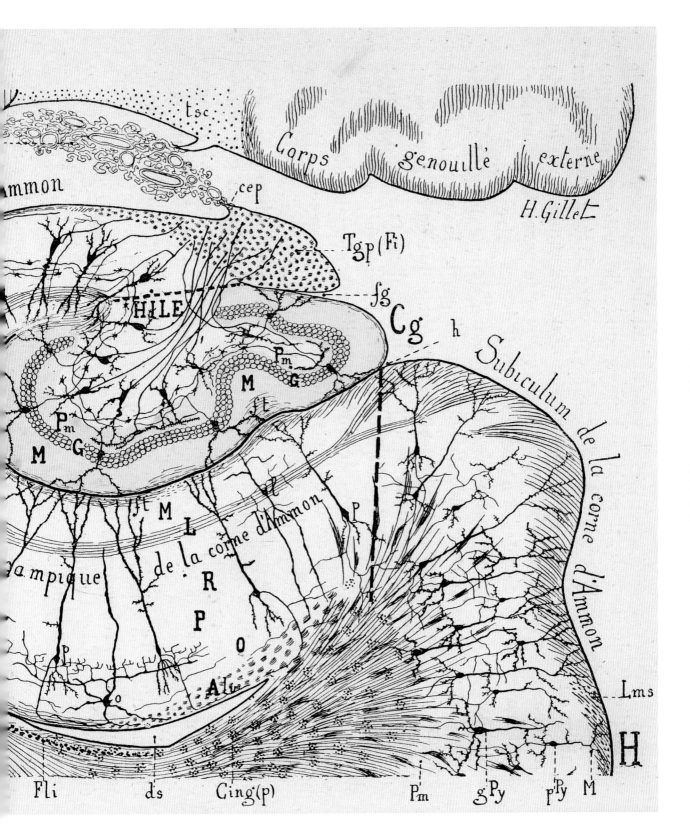

tsc

Corps genouillé externe

H. Gillet

mmon

cep

Tgp(Fi)

HILE

fg

Cg h

Subiculum de la corne d'Ammon

Pm

M G

Pm

M G

M L de la corne d'Ammon

ampique

R

P

P

O

Al

Lms

H

Fli   ds   Cing(p)   Pm   gPy   pPy   M

Olfactory bulb.
Camillo Golgi, 1875.

This 1875 drawing of a dog's olfactory bulb by
Camillo Golgi is but one of the many astonishing
architectures that were revealed by a staining
method that bears his name. Its application to
the study of nervous tissue marks the beginning
of modern neuroscience.

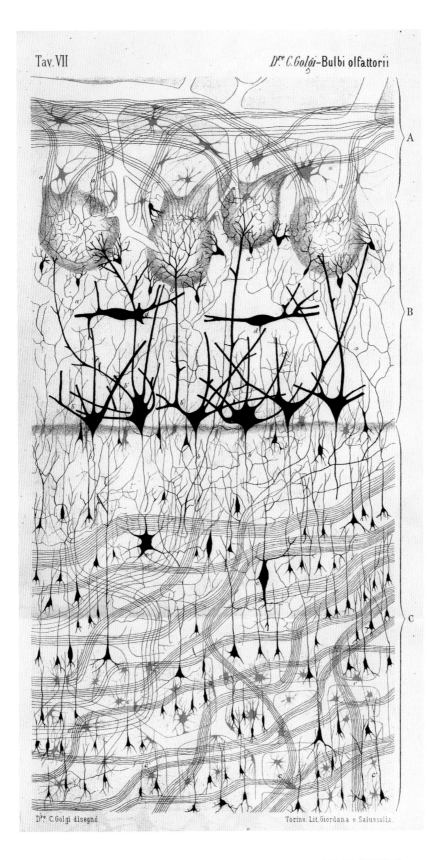

D^{re} C. Golgi disegnò   Torino. Lit. Giordana e Salussolia.

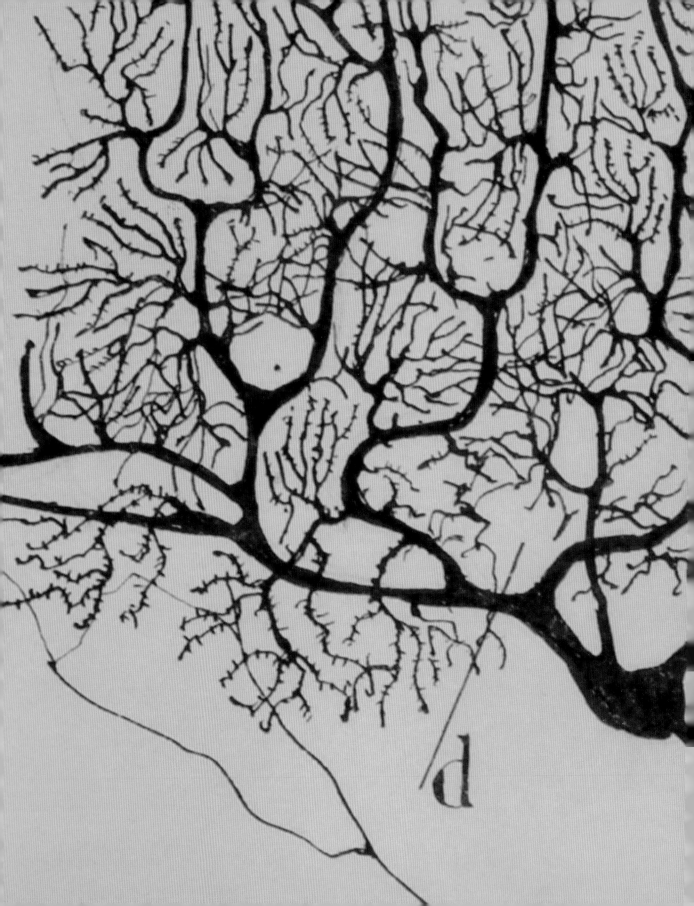

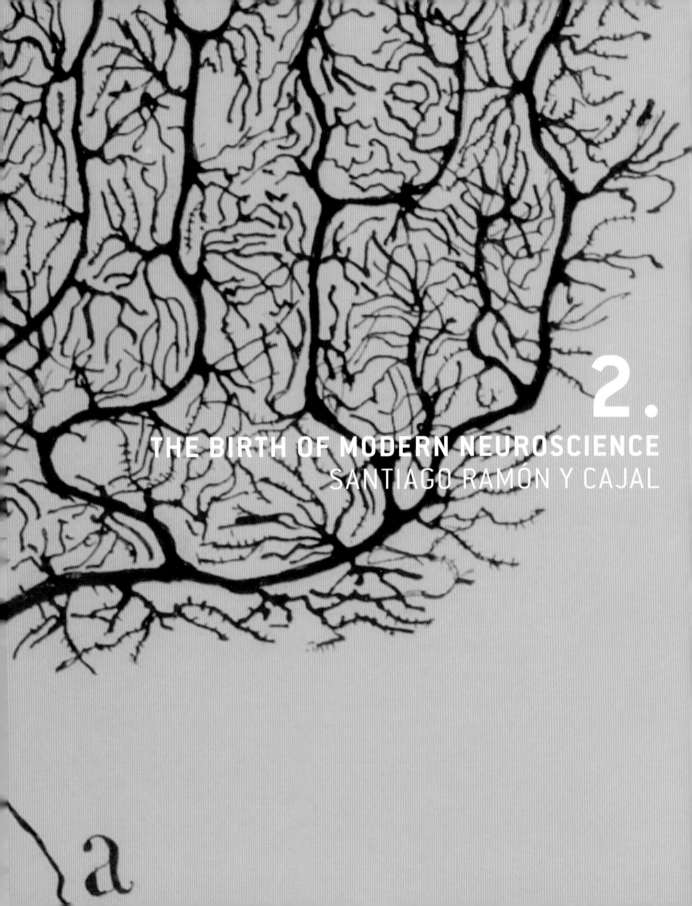

2.

THE BIRTH OF MODERN NEUROSCIENCE
SANTIAGO RAMÓN Y CAJAL

a

THE BIRTH OF MODERN NEUROSCIENCE:
SANTIAGO RAMÓN Y CAJAL

by Javier DeFelipe
Technical University of Madrid, Cajal Institute (CSIC)

The discovery of the *reazione nera* (black reaction) in 1873 by the Italian physician and scientist Camillo Golgi (1843–1926) set in motion a powerful research program that is still playing out today. Before then, our understanding of the microscopic structure of the nervous system was rudimentary. Since there were no techniques that enabled one to view neurons in their entirety, the brain was almost entirely unknown at the cellular level, making it impossible to study its patterns of connections. In fact, during the nineteenth century, the Reticular Theory, the most commonly held view about the organization of the nervous system, posited a diffuse network of nerve fibers fused with each other at various points—a vast, tangled net. When prepared with conventional techniques, tissue samples from the brain certainly appeared as a mess of intertwining threads, but Golgi discovered a novel way to treat brain matter, opening up an entirely new perspective. Using his new method, anatomists could, for the first time, visualize the entire span of neurons; with its widespread application, the foundational task of characterizing, classifying, and studying their connections began in earnest. Out of this revolutionary visualization method came a theory that did away with the reticular view, which was replaced with the hypothesis—still unchallenged to this day—that the fundamental organizational and functional units of the nervous system are individual cells. The establishment of this theory, called the Neuron Doctrine, marks the founding of modern neuroscience.

As is often the case in scientific research, it is one thing to discover an exciting new tool and quite another to fully take advantage of it; for more than a decade, few scientists bothered to even try the *Golgi method*. Fourteen years after its invention, it fell into the hands of the Spanish physician and scientist Santiago Ramón y Cajal (1852–1934). Cajal learned about the Golgi method in 1887 thanks to a fortuitous encounter with the psychiatrist and neurologist Luis Simarro (1851–1921). Cajal was so impressed by this powerful "new" method that he immediately applied it to study virtually the entire nervous system of several species. Meanwhile, he pondered why its discovery had failed so far to foment a scientific revolution:

> *I have already expressed above the surprise I felt when I saw with my own eyes the wonderful revelatory power of the chrome-silver reaction [Golgi method] and the indifference of the scientific community regarding this discovery. How could this disinterest be explained? Today, as I better understand the psychology of scholars, I find it very natural. In France, as in Germany, and more in the latter than in the former, a severe school discipline reigns. Out of respect for their master, it is common that disciples do not use research methods that have not been passed on by him. As for the great investigators, they would consider themselves dishonored if they worked with the methods of others.*[1]

The momentous day when Cajal first marveled at the revelatory power of the Golgi method is captured in several of his writings, here in his classic treatise, the *Histology of the Nervous System of Man and Vertebrates*:

Could the dream of such a technique truly become reality, in which the microscope becomes a scalpel and histology a fine [tool for] anatomical dissection?.... What an unexpected spectacle! On the perfectly translucent yellow background, sparse black filaments appeared that were smooth and thin or thorny and thick, as well as black triangular, stellate, or fusiform bodies! One would have thought that they were designs in Chinese ink on transparent Japanese paper. The eye was disconcerted, accustomed as it was to the inextricable network [observed] in the sections stained with carmine and hematoxylin where the indecision of the mind has to be reinforced by its capacity to criticize and interpret. Here everything was simple, clear and unconfused. It was no longer necessary to interpret the findings to verify that the cell has multiple branches covered with "frost," embracing an amazingly large space with their undulations. A slender fibre that originated from the cell elongated over enormous distances and suddenly opened out in a spray of innumerable sprouting fibres. A corpuscle confined to the surface of a ventricle where it sends out a shaft, which is branched at the surface of the [brain], and other cells [appeared] like comatulids or phalangidas.* The amazed eye could not be torn away from this contemplation. The technique that had been dreamed of is a reality! The metallic impregnation has unexpectedly achieved this fine dissection. This is the Golgi method![2]

A mere year after his meeting with Simarro, Cajal published his first significant article based on results obtained using Golgi's method. In this study, entitled "The Structure of the Nervous Centers of Birds," Cajal made two important contributions. He was the first to describe dendritic spines, the thousands of tiny protrusions that extend out of the branching arms of a neuron—structures that today still garner avid interest due to their noted plasticity and their function as the main targets for excitatory synapses in the cerebral cortex. Moreover, Cajal confirmed Golgi's conclusion that dendrites end freely and are unconnected to surrounding cells; however in contrast to Golgi, Cajal added that these properties also apply to axons and their branches. This entails that each neuron must be considered as a discrete unit, a single cell rather than merely a small part of a large mesh. With this conclusion, modern neuroscience was born.

✳ ✳ ✳

*Comatulids are marine invertebrates like sea lilies and feather stars. Phalangidas (or opiliones), also known as water harvestmen, are arachnids that superficially resemble true spiders but have small, oval-shaped bodies and long legs.

In Golgi and Cajal's day, most scientists employed the same microscopes and prepared their tissue using similar histological techniques. Microphotography was not yet established enough to record their anatomical findings, so they and their contemporaries mainly resorted to drawing what they saw under the microscope. This approach, while promoting close attention to detail, has severe limitations. First, the nervous system is an incredibly complex structure, and it can be difficult to distinguish artifacts of a tissue preparation from real elements of the tissue sample itself. Second, the observer is required, in the act of drawing, to highlight details he or she considers important; a key feature for one scientist can pass unnoticed by another, and two expert anatomists drawing from the same sample could potentially produce radically different diagrams.

An illustrative example can be found in the case of dendritic spines, which Cajal correctly conjectured to be anatomical manifestations of where two neurons connect. Today we know that they are implicated in fundamental mental faculties such as learning, memory, and cognition, but in the nineteenth century the jury was still out as to whether they even existed. To some contemporaries of Cajal, who used histological methods and microscopes more or less identical to his, these specks were nothing but an artifact, a "superficial precipitate, like a crystallization of needles, fortuitously deposited on the dendritic surface."[3] Although they undoubtedly saw them under the microscope, they did not include them in their figures. Cajal was able to prove that they weren't merely a quirk due to imperfections of the Golgi method only when he successfully visualized them using an unrelated staining method.

This episode displays the wide range of standards in precision and accuracy of scientific illustration among researchers; reflecting on this issue takes me back to my student days in the laboratory of Alfonso Fairén and Facundo Valverde at the Instituto Cajal in 1979. We were discussing the recent discovery (using the Golgi method) of a type of neuron called the "chandelier cell." It owes its name to the remarkable short vertical rows of candlestick-shaped boutons (microscopic swellings) that appear at the terminal portions of its *axon*, the neuron's long tail responsible for transmitting signals out to other neurons. We were curious why these cells had not already been described despite a century of intensive application of the Golgi method, and why suddenly everyone was stumbling upon them in their own samples. We joked that perhaps they were due to a freak genetic mutation that had spread in the 1970s. But the truth is that chandelier cells were only first "seen" so late because at times even the most talented scientists can have something literally under their eyes and yet fail to interpret or represent it accurately.

＊ ＊ ＊

In addition to being able to interpret a sample under the microscope, the scientist of the late nineteenth century was required to exhibit considerable artistic aptitude, as demonstrated by the magnificent drawings in the chapter that follows. When diagrams challenged dogma, questions about the scientist's hand inevitably arose, and this is almost certainly one of the reasons why Cajal's early studies passed unnoticed at first. Arthur Van Gehuchten (1861–1914) describes the historic moment during an 1889 conference when Cajal, who had been derided up until then, finally persuaded Albert von Kölliker (1817–1905), one of the most influential neuroscientists of the era, that his results were valid:

> The facts described [by Cajal] in his first publications were so strange that the histologists of the time ... received them with the greatest scepticism. The distrust was such that, at the anatomical congress held in Berlin in 1889, Cajal ... found himself alone, provoking around him only smiles of incredulity. I can still see him taking aside Kölliker, who was then the unquestioned master of German histology, and dragging him into a corner of the demonstration hall to show him under the microscope his admirable preparations, and to convince him at the same time of the reality of the facts which he claimed to have discovered. This demonstration was so decisive that a few months later the Würzbourg histologist [Kölliker] confirmed all the facts stated by Cajal.[4]

Kölliker was so enthusiastic about the conclusions that he later told Cajal:

> The results that you have obtained are so beautiful that I am planning to undertake a series of confirmatory studies immediately, adopting your technique. I have discovered you, and I wish to make my discovery known in Germany.[5]

Up until then, Cajal's drawings, no matter how accurate and virtuosic, had wielded little influence. It required the strength of one of the field's strongest and most respected voices to disseminate his observations and theories around the world—and this, only after direct examination of Cajal's samples under the microscope.

This early period of microscopy research presents a fascinating page in the history of neuroscience and biology as a whole. By forcing an improbable melding of scientific and artistic practices, it ignited questions that still ring today about the nature and pitfalls of visualization, interpretation, and representation in scientific research.

Pyramidal neuron.
Santiago Ramón y Cajal, 1899.

Modern neuroscience begins with the neuron, a cell responsible for gathering, processing, and transmitting information in the brain. This illustration shows a *pyramidal neuron* drawn from observation by Spanish neuroscientist Santiago Ramón y Cajal in 1899. Its nucleus and DNA reside in the *soma*, the thickened area at the center. What distinguishes a neuron from other cells in the body is the striking set of long appendages that radiate from it—think of them as antennae that neurons use to communicate with one another across the expanse of the brain. The thicker *dendrites*, the cell's "receivers," rise upward and outward in this illustration, while the thinner *axon*, the single "emitter" of a neuron, drawn here only in part because of its considerable length, shoots downward. Unlike with antennae, however, transmission in the brain is not wireless, and the infinitesimal spaces between these neural appendages gave rise to an epic scientific battle one century ago.

The hypothesis that a neuron could be discrete—distinct from yet connected to others—once divided the most credentialed voices in neuroscience, and the schism ultimately made its way to the august podium of the 1906 Nobel Prize awards ceremonies. The committee had chosen to honor both Cajal and his scientific rival, the eminent Italian Camillo Golgi, for their respective contributions to the study of the brain. Golgi believed the brain was continuous, made up not of individual parts but rather an uninterrupted reticulum, a single mesh. Cajal, on the other hand, argued that the existence of the neuron as a self-contained entity supported a view of the brain as a network of distinct, interconnected units. In their Nobel lectures, the two scientists laid out their respective visions of the neural world, while taking shots at each other—some of them courteous, others decidedly not.

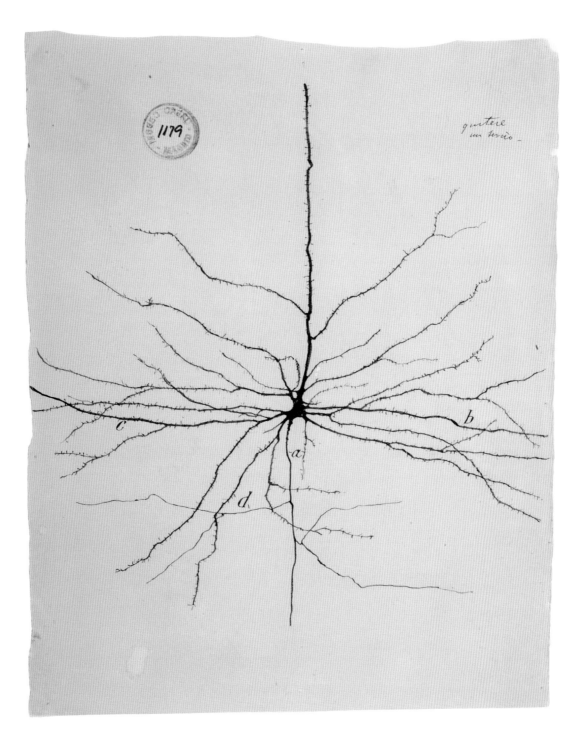

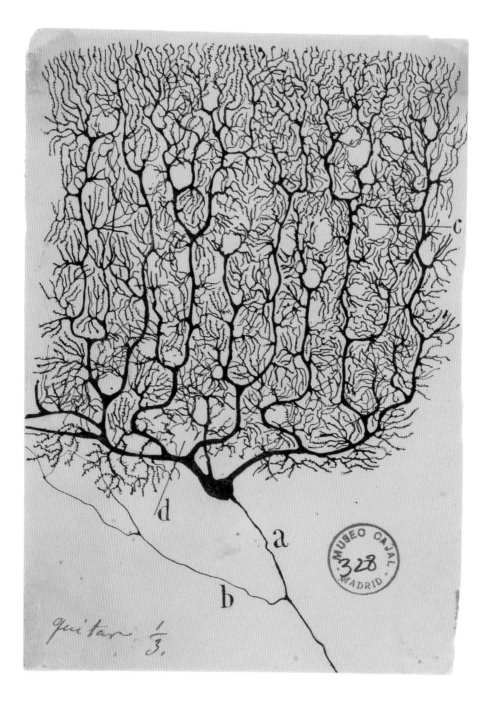

Purkinje neuron.
Santiago Ramón y Cajal, 1899.

By the time Golgi and Cajal's argument made
its way to the Nobel celebrations in Stockholm
in 1906, scientific opinion had mostly turned
against Golgi's theory, and the audience was
reportedly dumbfounded as he proceeded to
attack Cajal's views from the stage. That the
Spaniard won the neuron-versus-reticulum
debate, along with a host of other skirmishes,
is a testament to his prodigious knack for
synthesis. He apparently did not sketch
what he saw with a pen in one hand and the
microscope's focus knob in the other. Instead,
he is said to have drawn from memory, in the
afternoon, after a morning of observation.
Though the details were of course of great
importance, what mattered most to Cajal was
the general form, the common properties, the
essence of the specimen's overall architecture.
Studying tissues ranging from birds to humans,
he focused on the commonalities, uncovering
deep principles of brain organization. His amply
documented interest in the visual arts suggests
why his renditions of biological samples are
so exquisite and hints, perhaps, at why he was
able to identify underlying forms where others
saw only lines.

  This drawing shows a *Purkinje neuron*
that is strikingly different from its pyramidal
cousin on page 55 but built along similar lines.
Although the two cell types play very different
roles in the brain, they share the same blueprint:
at the center, the soma; radiating up and out, its
many thick, branching dendrites; traveling down
and beyond the frame, the long, thin axon.

Above: Letter containing the first description of reazione nera.
Camillo Golgi, 1873.

Though Cajal earned a choice spot in neuroscience history, and all manner of hagiography, his immortality is due in part to his archrival's discovery of a novel way of staining cells. The *Golgi method*, as it came to be called, was not a conceptual advance, but a procedural one; rather than offer new ideas, it provided a technique for manipulating brain tissue in order to reveal its structure under the microscope.

Golgi came upon his momentous discovery in 1873 in the bare-bones laboratory he'd set up in a kitchen; years later, Cajal performed initial experiments using the same staining method in his own private lab. Named the *reazione nera* (black reaction), Golgi's method allows a proficient user to stain individual neurons, and,

crucially, targets only a small percentage of neurons in the tissue, leaving all others invisible. Why some neurons incorporate the black stain— a mix of potassium dichromate and silver nitrate—while most do not remains a mystery to this day, but this selective viewing facilitated previously unthinkable access to the neural universe. If an anatomist were to peer at a slice of untreated gray matter under the microscope, it would appear as just that: more or less undifferentiated tissue in which neural structures would be entirely inaccessible to the eye. But let's suppose instead that the anatomist stained all the neurons within that tissue. The resulting tangle of axons and dendrites would be so dense that it would be impossible to distinguish one neuron's appendages from another's; instead of all gray, it would appear as all black. Identifying individual neurons would be like trying to pick out the arbors of individual trees from a helicopter flying above a dense forest: If all of them are visible, the tangle is simply impenetrable. The Golgi method provides a clever middle ground between seeing all gray or all black, marking only a very small number of cells and so affording the careful anatomist a view of whole individual neurons in all their glory.

The *reazione nera*'s revolution in visualization unleashed a revolution in thought and instituted a mode of discovery that continues today, in which novel techniques yield new insights. As Golgi excitedly reported in a letter to a friend (opposite), his new staining method would reveal the structures in the brain "even to the blind."[6] What the letter didn't anticipate was how difficult it would be to interpret its revelations.

A closer look at the dendrites, the "receivers" of a neuron, reveals hundreds of tiny notches, like specks of dirt along the edges—or so thought Golgi when he peered at the dendrites of certain neurons under his lens. He perceived this "fuzz" on the sides as nothing but an artifact, perhaps some stray silver that had failed to wash out in the final steps of the *reazione nera*. Cajal, on the other hand, believed that these specks could not be ignored, and in fact might be fundamental building blocks of the neuron. Thus Golgi drew smooth dendrites, whereas Cajal, looking at an identically prepared slide, rendered the notches—a striking reminder of the considerable role of interpretation in any empirical enterprise.

As they are reminiscent of the thorns on a rose, Cajal called these small protrusions *espinas* ("thorns" in Spanish), a charming bit of neuro-jargon that gets lost in the currently used term *dendritic spine*. Cajal conjectured rightly, as was proved decades later, that these spines must serve as points of contact between neurons. This illustration shows closeups of four dendrites with their many spines.

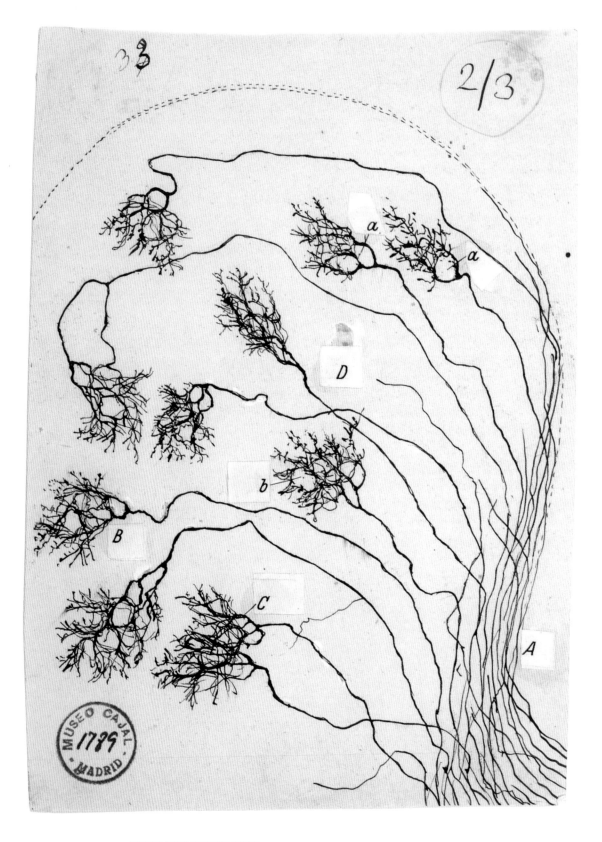

Opposite: Axons in the thalamus.
Santiago Ramón y Cajal, 1900.
Right: Axons wrapped around a neuron.
Santiago Ramón y Cajal, 1914.

Opposite the dendrites, an axon (the "emitter")
exits the soma to embark on a sometimes
improbably long journey to broadcast the
neuron's message. The muscles in our toes,
for instance, are controlled by neurons whose
somata are found in the spinal cord, implying
that the distances spanned by these single
cells can sometimes reach tens of inches or
even a few feet. In the illustration opposite,
each continuous line represents one neuron's
axon—whose faraway soma and dendrites are
not shown. Here the axons enter at the bottom
right and form a highly branched web at their
extremities, a pattern specific to a region of the
thalamus. Not drawn but strongly implied by
this effusive branching is the focus of all this
attention: the neurons in the thalamus to
which the axons have traveled to deliver their
messages. This points us to an interesting and
important fact: Each part of the brain bears its
own signature architecture of axons.

The illustration at right zooms in a bit
to show axons (represented by thin lines)
converging and snaking around a neuron's
soma (the thickened area to right of center) and
dendrites (the thick lines emanating out of the
soma). The axons are now close enough to make
physical contact with the dendrites of adjacent
neurons, which is shown in the next illustration.

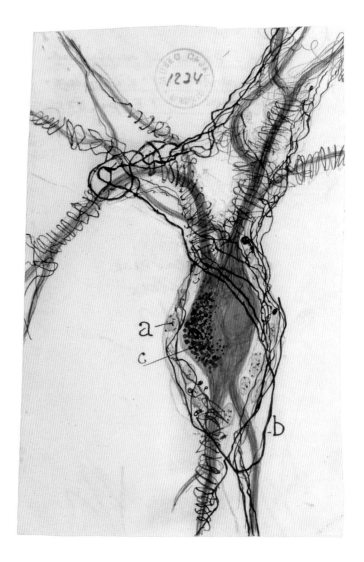

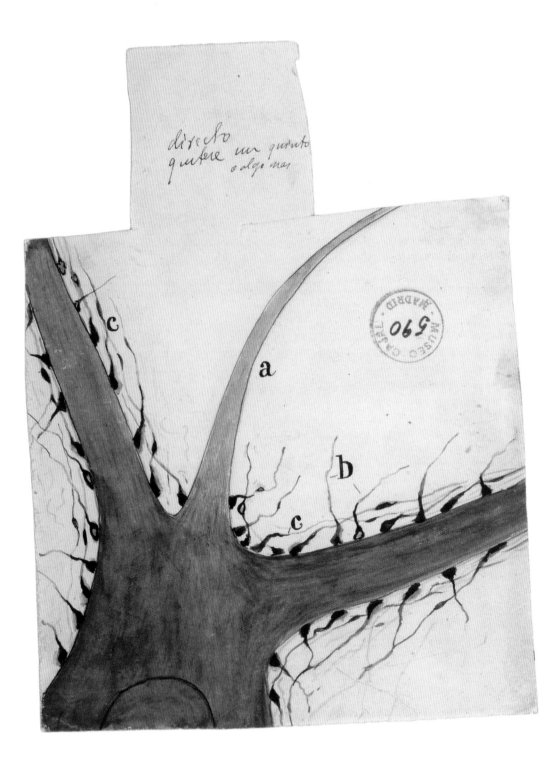

PORTRAITS OF THE MIND

Boutons forming synapses with a neuron.
Santiago Ramón y Cajal, 1903.

The relationship between axons and dendrites leads us, inevitably, to the synapse. The dendritic spines shown on page 59 form one half of the connection between neurons; their counterparts, *synaptic boutons* (one of which is labeled "b" here), bud from an axon and are attached to spines. (Note that in this example showing a spineless neuron, the boutons are connected directly to its soma and dendrites.) Together, dendrite and bouton form a synapse, the elemental means of communication between two neurons. Each neuron broadcasts its message to up to thousands of "listening" neurons via the synapses that have formed between them. By and large, this conversation is unidirectional: Information travels from the soma of the broadcasting neuron down its axon to its many boutons. There it crosses into the spines of the listening neurons it has formed synapses with. Once it has crossed, it propagates down their respective dendrites all the way to their somata, which in turn send the information they've collected to the neurons *they* broadcast to. And if this weren't complicated enough, spines are not always players in this story. As shown in this drawing, boutons (each one sent from a distant neuron) also form synapses directly onto the main shaft of dendrites, the two thick extensions radiating out of the soma (bottom left), and even with somata themselves.

Network: cerebellum.
Santiago Ramón y Cajal, 1904.

Now we may gather all the pieces: a variety of neuron types (like those on pages 55 and 56); dendrites for listening; axons for talking; synapses acting as the bridge between them. Together these elements form a network of neurons, or a circuit. If you wish to understand how a radio set works, your first step would be to crack open the box to see how it's wired up. This same premise fueled Cajal's investigations of the nervous system, and it drives many laboratories to this day: Understand how a circuit is laid out and you will uncover clues about how it works. After Cajal, the story of neuroscience became, fundamentally, one of wiring.

This drawing shows one example of a circuit with numerous neurons arranged together to process information across the cerebellum. Each neuron has its own determined place and role in the circuit, and the Purkinje neuron we encountered earlier is only one of many in this network. The next five drawings depict some of the brain's circuits mapped out by Cajal.

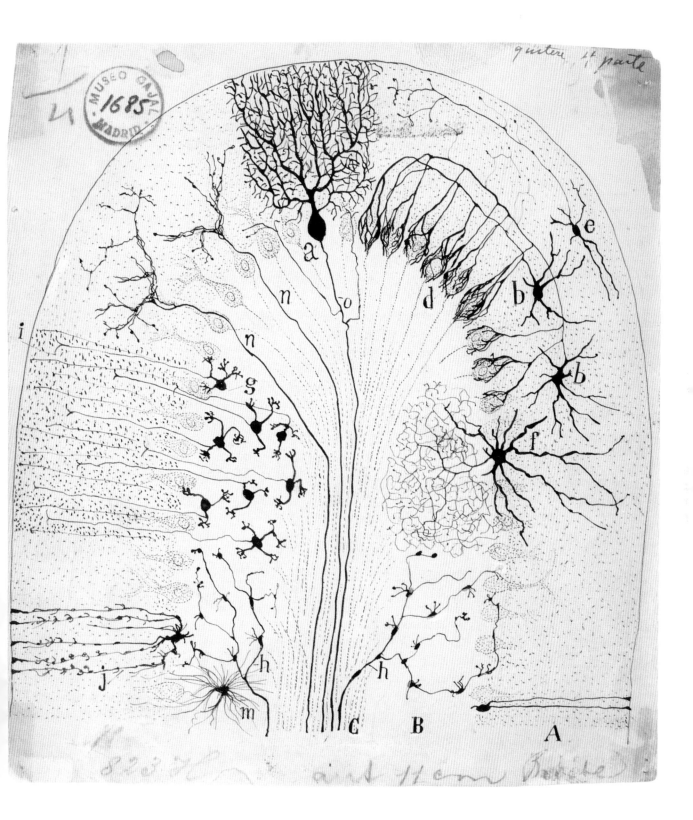

Network: neocortex.
Santiago Ramón y Cajal, 1899.

Cajal's circuits did not appear under the microscope as they do in his diagrams, of course:
Cell types were not conveniently clustered and
isolated for clarity. Many would have overlapped
in the stained nervous tissue, making it difficult
to parse. Only through patient examination of
countless slides over the course of years was
Cajal able to extract each neuron from the
messy background and synthesize his findings
on the page.

　　This drawing shows superficial layers of
the *neocortex*, the outermost part of the brain
and, in humans, the one that forms the
folds on its surface. As one's eye travels downward from the top of the illustration, one descends
deeper into the brain. The pyramidal neuron
labeled "E," whose soma lies comfortably deep,
sends its axon straight down (and out of
the drawing) and sends a long, thick dendrite
upward, fanning out at the top—the better to
catch incoming information from the dense
bundles of axons (not depicted) coursing from
far and wide across the brain.

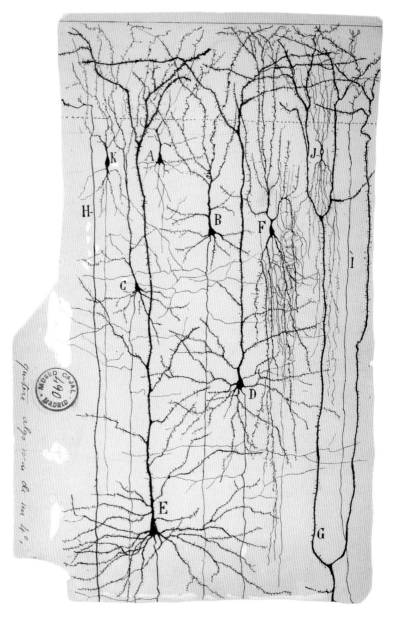

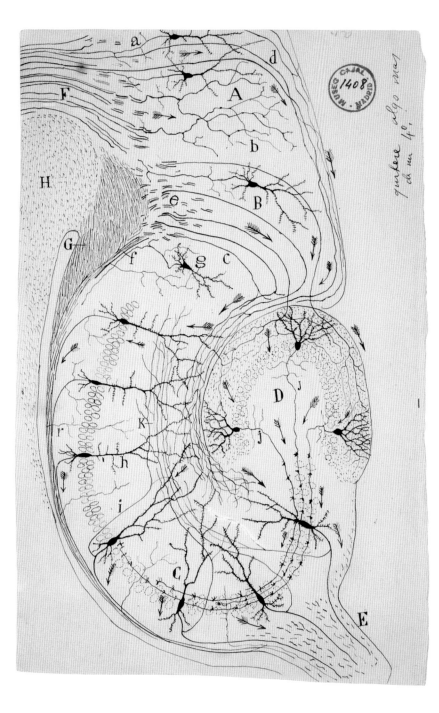

Network: hippocampus.
Santiago Ramón y Cajal, 1901.

Here is an early diagram of the hippocampus, so named because of its vague resemblance to a sea horse (*hippo* means "horse" in Greek; *kampos* means "sea monster"). If the hippocampus is removed or damaged, crucial abilities such as memory are impaired. London cab drivers, noted for their geographic prowess, were recently discovered to have peculiarly shaped hippocampi; this complements the finding in rodents that neurons in this structure encode a spatial map of their environment. Moreover, the hippocampus has proved to be a reliable site for neural plasticity, a key property of any system that must learn and adapt to its surroundings. These properties and many others have made the hippocampus a tantalizing brain area to a legion of laboratories investigating learning and memory. There remain many unknowns about what the hippocampus does and how it works, but Cajal provided the canonical map, along with the arrows, that guide our explorations to this day. Compare this drawing with Dejerine's representation on page 45 and notice the marked improvement in precision and detail.

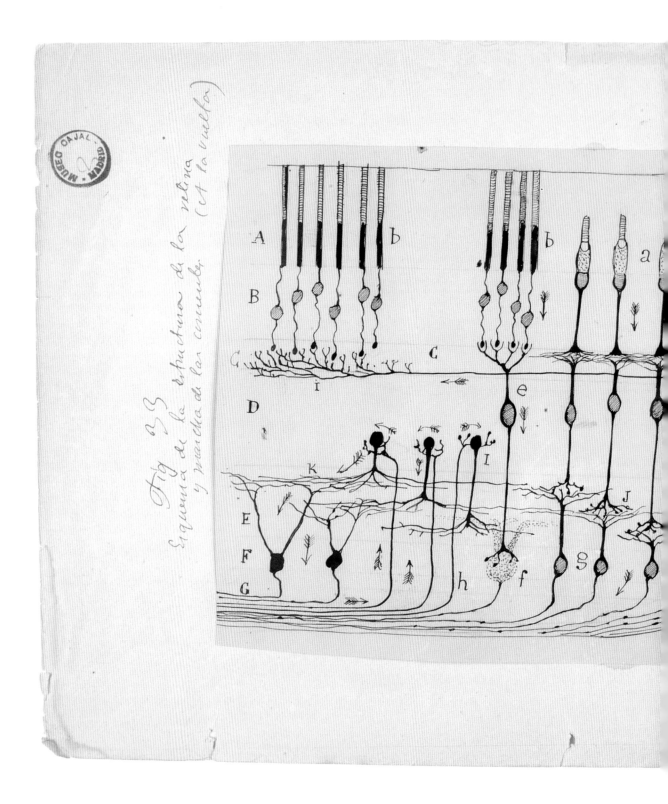

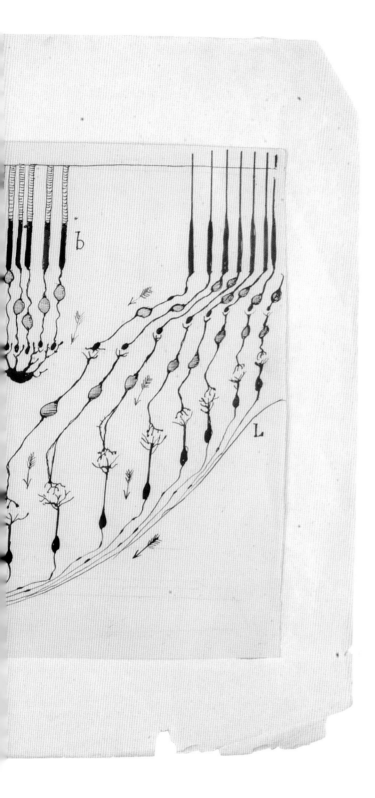

Left: Network: retina.
Overleaf: Network: visual system.
Santiago Ramón y Cajal, 1901.

The drawing at left shows the circuitry of the eye's retina along with small arrows depicting the flow of information within it. The critical reader might wonder how Cajal, with only dead brain tissue at his disposal, was able to conclude that information travels in a certain direction: into the dendrites and soma, then broadcast to the next set of neurons via the axon.

This strict directionality is one of those scientific facts that we take for granted today (with interesting exceptions), but it was neither obvious nor a fact until Cajal made it so through a simple but fruitful series of observations. In order for the retina to convey information to the brain, he surmised, the eye must send it across wires of some kind, either axons or dendrites. And indeed, he discovered that while dendrites are relatively short and confined to the retina (labeled A in the lower left corner of the drawing on the following spread), long axons (labeled B) shoot out and fan across large expanses of the brain, carrying visual information to the other areas depicted in the drawing (labeled D–P). Using this reasoning, he filled in the one-way streets of the neural map.

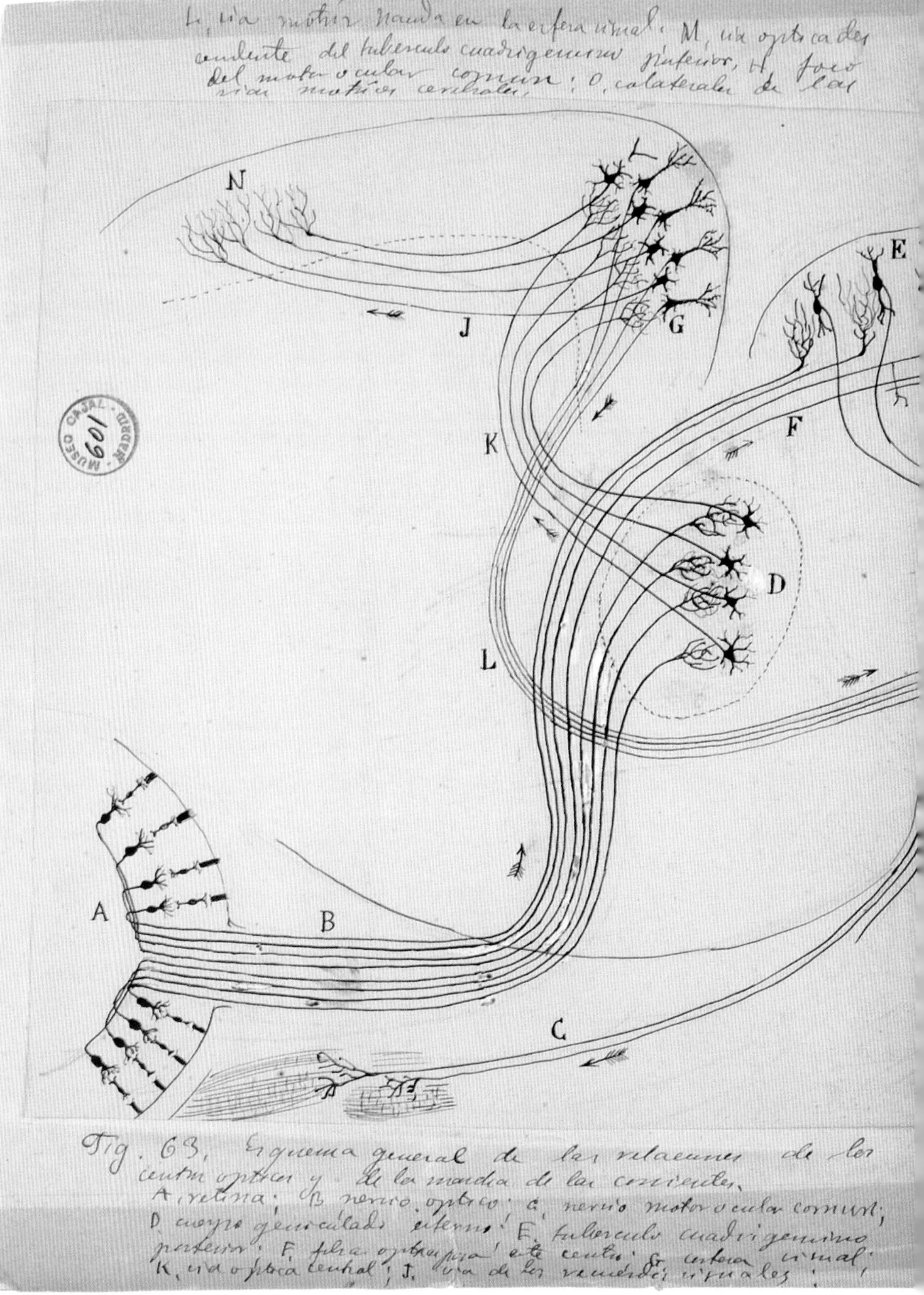

L, vía motriz nacida en la esfera visual; M, vía optica descendente del tuberculo cuadrigemino posterior, H foco del motor ocular comun; O, colaterales de las vías motrices centrales.

Fig. 63. Esquema general de las relaciones de los centros opticos y de la marcha de las corrientes. A retina; B nervio optico; C. nervio motor ocular comun; D cuerpo genculado externo; E. tuberculo cuadrigemino posterior; F fibra optropsia; este centro de corteza visual; K vía optica central; J. vía de los movimientos iguales;

PORTRAITS OF THE MIND

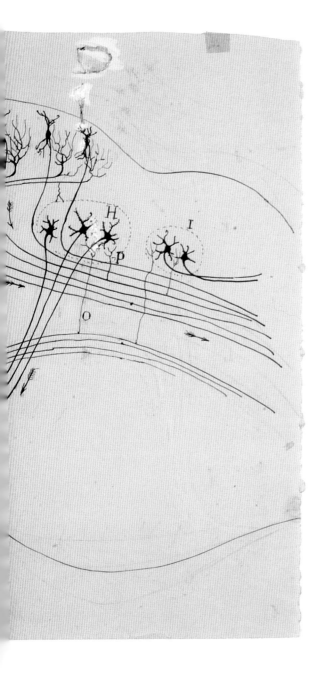

Photomicrographs of Golgi-stained tissue.
Santiago Ramón y Cajal, 1918.

Lest the clarity of Cajal's vision belie the complexity of his subject, this image shows a representative sample of his raw data: photographs of a slide he prepared using the Golgi method. It suggests the daunting task he faced when setting out to make sense of the circuits in the brain. While the staining method cuts out most of the information by staining only a small number of neurons, there remains nonetheless a confusing tangle of wires in the background, one that beckoned to a researcher like Cajal and required decades of painstaking microscopic study to decode. In fact we are still at pains to make sense of it. Golgi's *reazione nera* provided the first means to untangle the circuitry, and Cajal ushered in an approach to studying it that continues to bear fruit. The next chapters will explore where Cajal's legacy stands today—a diverse wealth of chemical, genetic, and physiological means to ignore the mess and focus on small, isolated parts that enable scientists to piece together the whole.

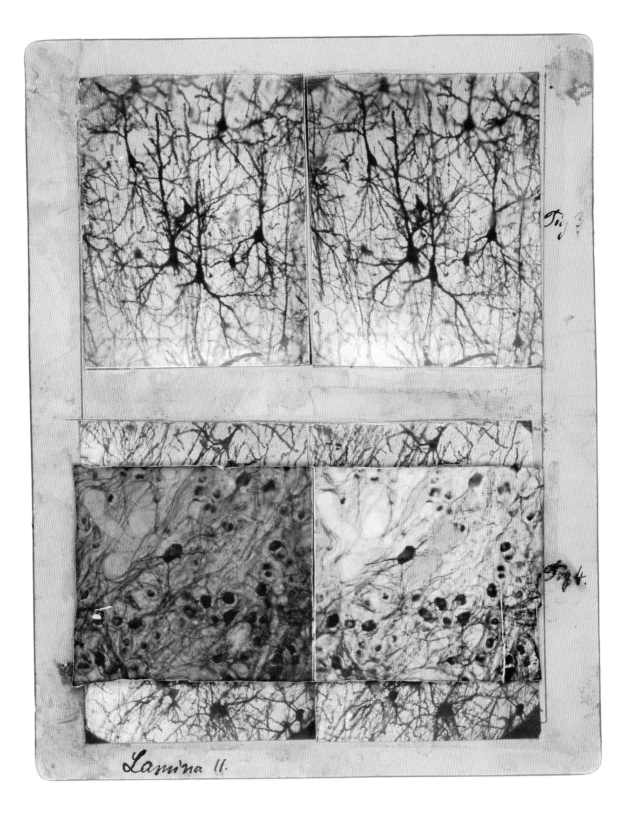

Lamina II.

Fig. 3

Fig. 4

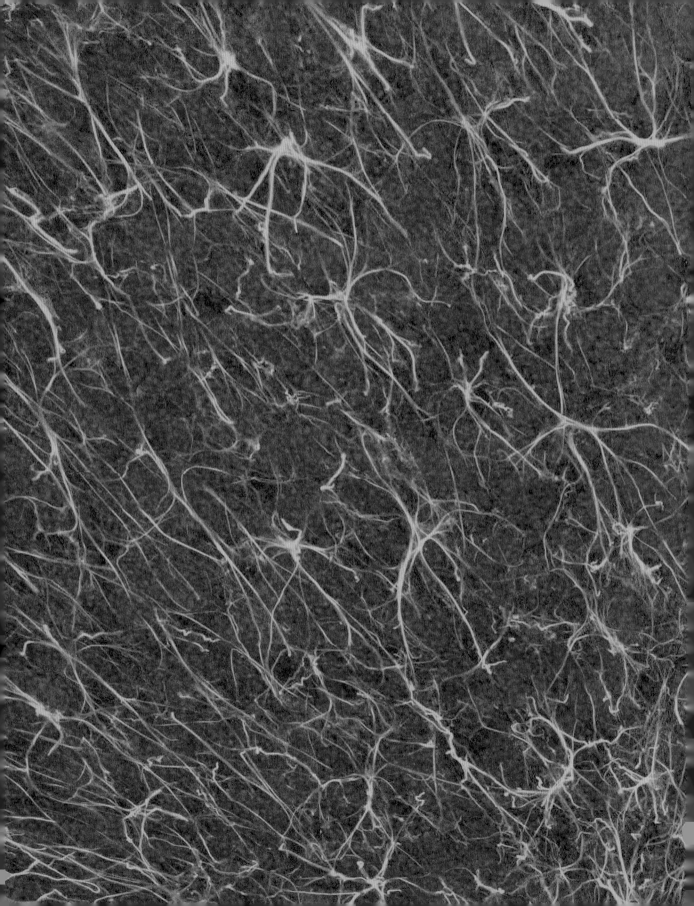

3.

AFTER CAJAL
FROM BLACK AND WHITE TO COLOR

AFTER CAJAL: FROM BLACK AND WHITE TO COLOR

by Joshua R. Sanes
Harvard University

In the decades following Cajal's first application of Golgi's black reaction, many neuroscientists used it and its various successors to begin filling in the innumerable remaining blanks, yet it is fair to say that progress in mapping the brain's circuits slowed over this period. One reason is that Cajal's skills, brilliance, and energy were unparalleled; he was clearly gifted with the ability to intuit cellular and circuit principles from fragmentary evidence. In addition, though, there are intrinsic limitations to the Golgi method. Three have proven particularly crippling.

First, to understand the function of neural circuits, it is necessary to record the electrical activity of particular neurons or even manipulate their activity while monitoring a subject's behavior. This, of course, requires an ability to visualize neurons in live animals, but the black reaction can be applied only to tissue that has been taken from a euthanized animal, treated with harsh chemicals, and chopped into fine pieces. By the time Cajal saw his neurons, they were long since dead.

Second, we will never understand the workings of a machine with a hundred billion elements by studying them one at a time. We need to classify the brain's neurons into subtypes that share structural and functional features, then generalize from the individual to the class. For example, neurons that enable us to see share many characteristics with those that enable us to hear, but there may also be systematic differences between them. Even within the visual system, neurons that detect objects moving upward may be fundamentally different from those that detect objects moving in the opposite direction. The black reaction stains only a few of the neurons in a particular sample, more or less at random. In other words, it doesn't mean anything that certain cells are made visible by the Golgi stain; they are not related in any particular way, so any attempt at categorization must therefore be done post hoc, based on comparing numerous samples, and without any knowledge of how the cell in question might have functioned.

Third, tracing circuits to determine how this machine works requires visualization of many neurons and the connections among them. Yet, the black reaction is only useful when it labels neurons sparsely: As more and more neurons are made visible, it becomes more difficult to distinguish between them and ultimately impossible to distinguish them at all. Cajal actually intuited circuitry by assembling in his mind the images of neurons labeled in isolation throughout consecutive sections of brain tissue. Needless to say, this approach is error-prone.

※ ※ ※

An unlikely solution to the first two problems was provided by *Aequoria victoria*, a nearly transparent jellyfish that floats in the Pacific Ocean off the shores of Washington state. The improbable path from Pacific to brain transformed neuroscience while providing a wonderful object lesson in the power of nonapplied, curiosity-driven research to catalyze quantum leaps that could never have been anticipated or planned by goal-oriented programs. This body of work won a Nobel Prize in 2008 for its three contributors—Osamu Shimomura, Martin Chalfie, and Roger Tsien—each of whom followed his own lodestar but, as a group, ended up fomenting a revolution in the study of the brain.

Like some other species of jellyfish, *A. victoria* is bioluminescent—it emits a greenish glow. In the 1960s, Shimomura, a biochemist, searched for the basis of this bioluminescence, eventually finding a pair of proteins that did the trick. One glowed blue when exposed to the calcium salts in seawater; the other emitted green light when in the presence of the blue-glowing one. Shimomura named the blue protein *aequorin*, after the jellyfish, but his imagination apparently ran dry thereafter; he named the second one *Green Protein*, later expanded to *Green Fluorescent Protein* and abbreviated as GFP. For some years, aequorin was the valued product of this work: By monitoring its blue light, scientists could detect low levels of calcium salts, which turns out to be quite a useful thing to do.

GFP lay dormant until the 1990s when a man named Douglas Prasher determined the DNA sequence that encoded GFP. Cells are able to read this sequence of the four nucleotide bases that make up DNA—As, Ts, Cs and Gs—and use it as a template to generate the sequence of amino acids—leucines, lysines, valines, and the like—that make up the green-glowing protein. Martin Chalfie, made the first link between GFP and neuroscience, by inserting the piece of DNA that "spelled out" GFP into neurons in such a way that it led the cells to manufacture their own GFP. When he illuminated these cells with a blue light, they glowed green! Among the cells into which he inserted GFP were neurons, initially those of a lowly worm. In his first pictures of GFP-containing worms, the basic physical structures of the neurons—axons, somata, and dendrites—were brightly outlined against a black background.

The considerable value of this weird bit of chimerism was immediately apparent to many neuroscientists, including me. My colleagues and I had been trying to find ways to mark specific types of neurons in mice, so that we could follow their development, identify the synapses they made, and ask what went wrong with them in animal models of neurological diseases. A general method that allowed us to insert any desired gene into the genome of a "transgenic" mouse had already been devised to help us accomplish this goal. Why not use the GFP gene? When my colleague, Guoping Feng, did just this (after a lot of trial and error), and then illuminated slices of the mouse's brain, its neurons lit up. We were specifically interested in motor neurons (the ones that send axons to muscles) and retinal neurons (the ones that send axons from the eyes to the brain), but around the same time many others used a similar strategy to mark their own favorite cells. Pretty soon, the world of neuroscience was full of researchers shining blue lights at transgenic mice, fish, flies, and worms.

※ ※ ※

So what's the big deal? Why isn't this just a green version of the Golgi stain? One key is that GFP allows scientists to study neurons in *living* animals and tissues. Once the gene has been successfully and appropriately inserted into the genome of a live mouse (a technically complex endeavor, but one that had become routine by the early 1990s), the rest is automatic: the DNA makes RNA, the RNA makes protein, and the protein glows when illuminated. Indeed, my colleague Jeff Lichtman quickly applied to GFP methods he had already developed for inferior dyes, making "movies" of synapses forming in live mice—something that had never been seen before. We also used this method to look at disease progression in a mouse model of an invariably fatal disease called amyotrophic lateral sclerosis, also known as Lou Gehrig's disease.

With the full arsenal of molecular biology at our disposal we had the ability to switch on the gene for GFP only in particular cells of interest (see page 81, middle column). In some mice, we labeled random assortments of cells—more or less as Cajal had done a century prior. Since we could do so in an intact, living brain, we could follow them over time and ask, for instance, whether they formed new synapses when the animal was exposed to a new environment. In other mice, we labeled particular cell types, which could then be identified and studied as a class. Rather than looking at random groups of cells as Cajal had done, we were looking at targeted groups of similar cells. For the first time, the stain *meant* something: if they were lit up, then they somehow all belonged to the same "family." For example, returning to our initial interest in the neurons that connect the eye to the brain, we found one small subset of these cells—perhaps 5 percent of the total—that seem to "point" in one direction (see pages 84–85). Remarkably, these same cells tell the mouse about objects that move in that direction—a relationship of structure to function that, I like to think, would have warmed Cajal's heart. Indeed, a main thrust of my lab's current work is to mark retinal cells that tell the brain about specific features of the visual world, so we can trace the circuits that endow them with their properties and find out how they wire up during development.

But what about the third limitation of the Golgi stain—that once too many cells are made visible, they can't be distinguished from one another? In that regard, the green stain was perhaps more aesthetically pleasing than the black one, but fundamentally no better. The third Nobel-winning scientist in the GFP saga, Roger Tsien, pointed the way to a solution to this problem. Tsien tinkered with the gene for GFP to change the color of the light that it emitted—yellow or blue or red, for example, instead of green. He and other like-minded chemists also started searching and finding fluorescent proteins in other sea-dwelling creatures, such as corals. Guoping Feng and I (and, again, a host of others) immediately saw that these proteins—which we called, as a group, XFPs—could be combined to distinguish some cells from others. Imagine, for example, one male mouse configured to express the green-glowing GFP in one random subset of neurons, and a female mouse configured to express a red fluorescent protein, RFP, in another random subset. Now imagine a marriage of these two. By the laws of genetics, some of their offspring would inherit both the GFP and the RFP genes, thus their brains would exhibit three colors of neurons—some green, some red, and some with both GFP and RFP, which would appear yellow. Now we had a way to distinguish individual cells in more densely visualized samples.

Manipulations like this broke the tyranny of monochrome that had plagued Cajal, but were not, in fact, much of a triumph. To map the billions of neurons in a brain, or even the tens of thousands in any little nook or cranny of the brain, three colors is scarcely a drop in the bucket. Enter Jean Livet, a young French scientist who had begun his career using GFP-marked mice to study the development of motor circuits and envisioned a way to mark many different neurons with many different fluorescent colors. He then harnessed the imaging facilities in Jeff Lichtman's laboratory, and the molecular genetic facilities in mine, to implement his creative scheme. Jeff also contributed its evocative name—Brainbow.

What Livet invented was a kind of molecular slot machine. He created DNA molecules in which the sequences encoding red, blue, and green fluorescent proteins were cleverly linked together in such a way that once any one color was turned on, the others would be turned off. He then generated transgenic mice in which approximately ten copies of these linked sequences were inserted into a chromosome, one after the other. The switch that allowed them to be turned on and off is called Cre recombinase, nicknamed "Cre," a protein derived from yeast. Cre, which is also inserted into the transgenic mice, is equally likely to turn on red or blue or green—and thereby turn off the other two colors—in each chunk of DNA. Cre makes each choice independently on each chunk and in each cell, just like a slot machine does when you pull the handle. Assuming the machine isn't rigged, the "fruit" that appears in the first position when the wheels stop spinning has no influence on which fruit appears in the second and third positions. And each time you pull the handle (analogous to Cre acting in one cell after another), a new combination of colors appears.

So what happens? Let's take just two colors and four copies of the sequence as an example. One neuron could end up with green-green-green-green, another green-green-green-red, another green-green-red-red, another green-red-red-red, and another red-red-red-red-red. They would look green, chartreuse, yellow, orange, and red, respectively. Five hues from two colors. Now imagine three colors and ten copies, and it is easy to see how neurons could end up with any one of more than one hundred colors. This strategy for obtaining many hues from a few basic colors is no different than the one implemented in television screens, which combine different intensities of red, green, and blue light to generate innumerable hues. What Jean did was to implant this color-generating machine into a mouse brain.

The multiplicity of hues allows us to distinguish a neuron from its near neighbors, as they each have their own color (usually). Thus, one can follow their extensions in the tangled, colorful *neuropil*, the area between the cell bodies of neurons where axons, dendrites, and the supporting cast of *glial cells* wind their way through and among each other in a vastly complex tangle (see opposite, bottom right square). Studies using the Brainbow method are only just getting under way, but it has already helped map connections in the neuromuscular system and in parts of the brain that process vision and audition.

As it stands, the biggest limitation of the Brainbow method is that different colors help distinguish one neuron from another, but they don't tell us what type of neuron either one is: the colors still have no meaning. We can follow the entangled extensions of multiple neurons, but we still cannot tell what they do from their colors. And because labeling is random, any particular type of neuron will probably have a different color in a different mouse. Wouldn't it be great if we could endow colors with meaning—marking, for example, neurons of one type in a range of reds and oranges, those of a second in yellows and green, and those of a third in blues and violets? We are thinking of ways to do just that. If and when we succeed, we will be able to visualize the brain in a way that reveals the magnificent order hidden within the vastly complex organization of its neurons.

Golgi GFP Brainbow

One cell

Several cells

Many cells

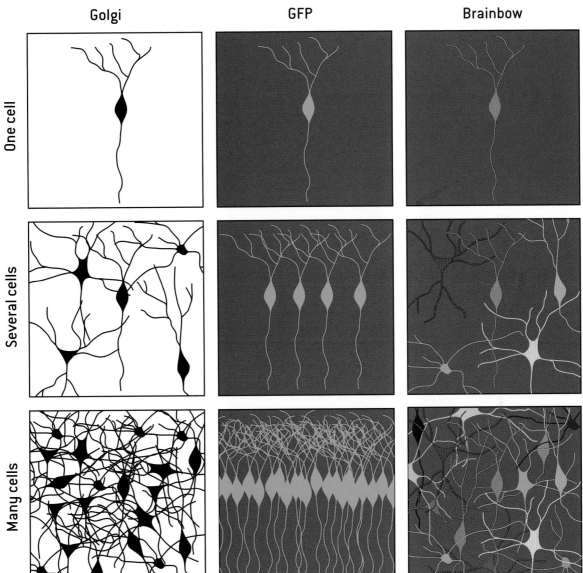

Opposite: Thy1.
Miguel Vaz Afonso and Jeff Lichtman, 2006.

The twentieth century has brought biomedical researchers a standard tool kit for cutting, pasting, and assembling genes and parts of genes—not unlike the way film editors of old put together filmstrips. The ability to shuffle pieces of DNA around with such ease has markedly increased our power to visualize and study neurons. Modern molecular biology's answer to the Golgi stain is a genetically engineered animal called the *Thy1 mouse*.

Using genetic cutting and pasting, researchers have created transgenic mice in which they associated a fluorescent protein with a gene called *Thy1*, which is used by most neurons. For reasons that remain unknown, only a very small percentage of the neurons in this genetically modified mouse light up, enabling researchers to distinguish between them. This image shows the intricate structure of those select few as they glow in the cerebral cortex of a *Thy1* mouse observed under a microscope. Once these shining, isolated neurons are made to stand out from the great tangle of brain matter left in the dark, they can be easily distinguished from one another and studied. Cajal would undoubtedly have marveled at this animal— a living, breathing Golgi stain.

Overleaf: JAM-B.
In-Jung Kim and Joshua Sanes, 2008.

All the cells in an organism share the same genome, but each cell uses only a certain portion of the genes contained within it. These genes provide a cell with its particular flavor and abilities: For example, genes specific to liver function are said to be "switched on" in liver cells but not in kidney, heart, or hair cells. Nerve cells are no exception, and each type of neuron switches on a set of genes that defines its character.

The terrific advantage of using genetics instead of traditional chemical stains is that the scientist can interact with—and thereby study—neurons by employing the very chemical language of cells, in a sense hijacking their instructions for research purposes. With genetics, in addition to fluorescently marking and visualizing neurons of interest, one may also measure, and even manipulate, their activity by introducing or removing genes that enable the recording or perturbation of neural function.

This example shows the small subset of neurons in which the fact that a gene, *JAM-B*, was switched on was used, in turn, to switch on a fluorescent protein. In the vast majority of neurons in the retina, *JAM-B* and its fluorescent follower are off so those cells are left in darkness, no longer confusing the picture. With only the selected cells in view, a striking fact immediately reveals itself in this photograph: All of their dendrites are aligned in the same direction, giving the cells the appearance of a set of arrows pointing upward. Moreover, these neurons exhibit a common physiological property in that they detect only those objects moving in an upward motion. While their purpose and detailed function still remain under investigation, this convergence of molecular biology, anatomy, and physiology is a template for neuroscience in the twenty-first century.

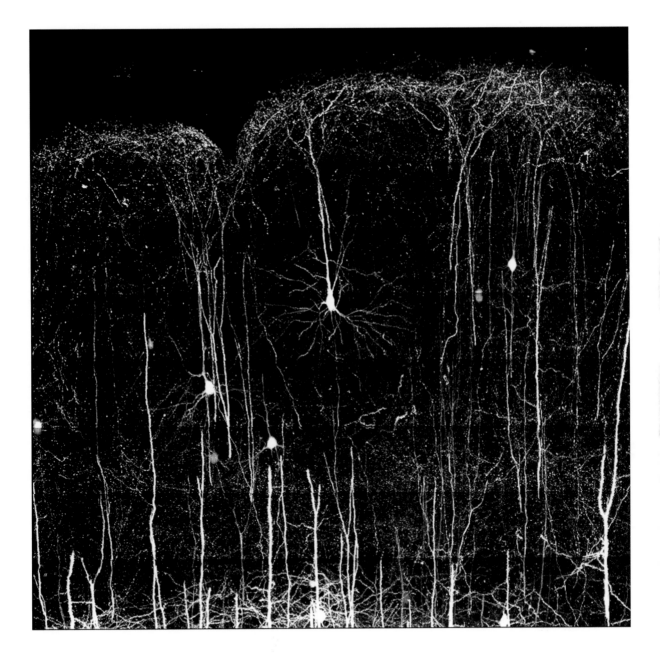

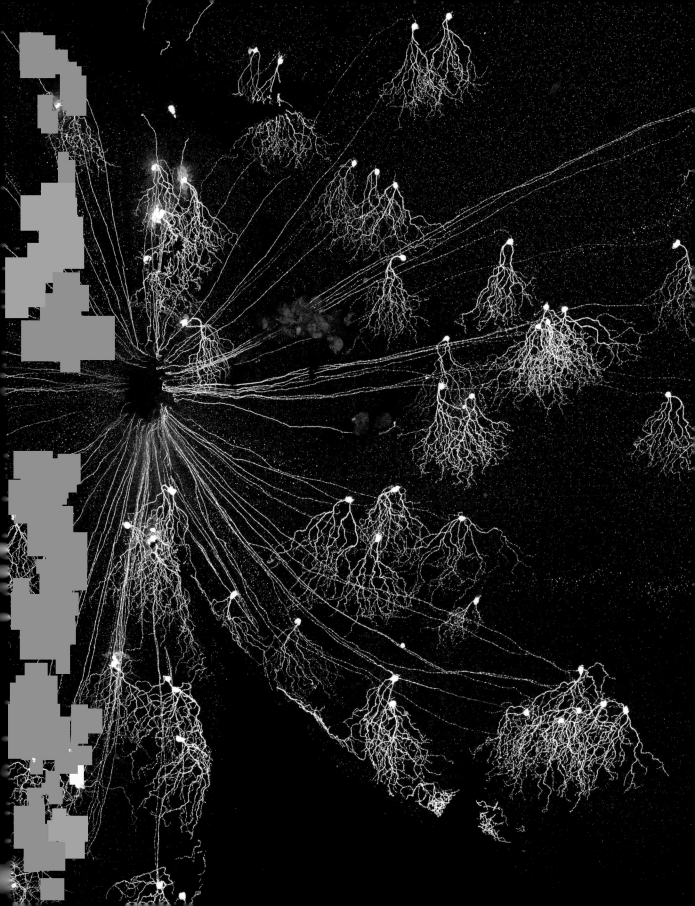

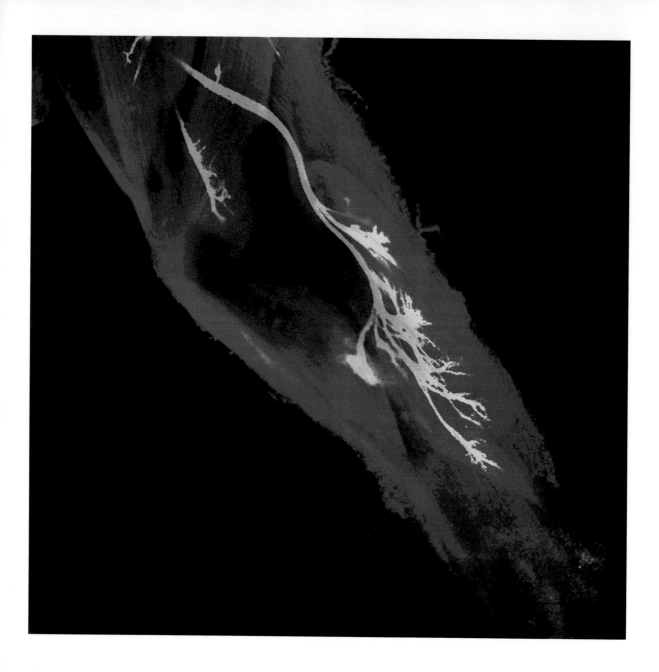

PORTRAITS OF THE MIND

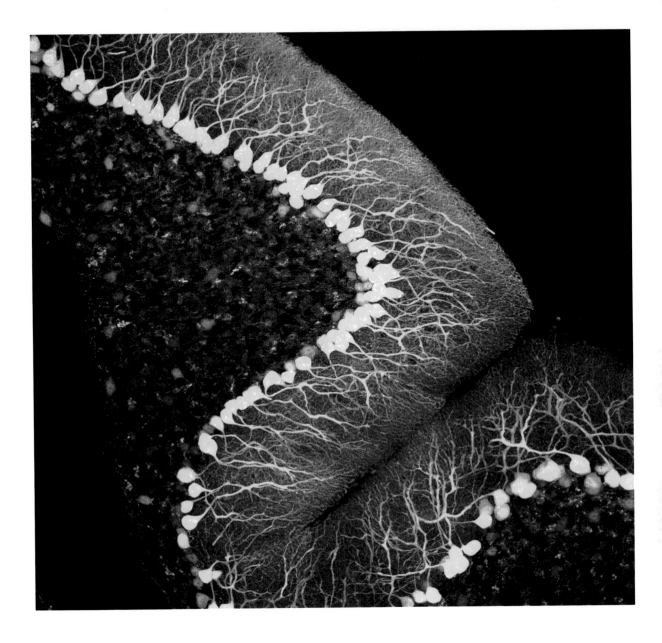

Opposite: Motor neurons in the forelimb of a mouse.
Gulsen Surmeli and Thomas Jessell, 2008.
Above: Cerebellar Purkinje neurons.
Aric Agmon, 2003.

Neuroscience has quickly embraced this power-ful approach to selecting subsets of neurons. It enables an ever more precise study of the circuits in the nervous system, breaking each area into its constitutive parts. In the image opposite, researchers have illuminated only the axons of those neurons (shown in green and yellow) that control the contraction of muscles (red) in a mouse's forelimb. Using this specific marking strategy, they can gain insight into how the illuminated cells mature over the course of life by studying them at different periods in their development.

The image above shows a portion of the cerebellum in which only its Purkinje cells (the same type as the one drawn by Cajal on page 56) are lit up (they are pseudocolored in yellow here).

Green Fluorescent Protein.

Nathan Shaner, Paul Steinbach, and
Roger Tsien, 2004.

The gamesmanship described above would have been impossible without a family of markers that have become ubiquitous across the whole of biomedical research. The one that begat them all is a molecule found in a strain of jellyfish that exhibits a striking fluorescent glow. Thousands of ground-up jellyfish later, a single glimmering molecule was isolated and named *Green Fluorescent Protein*, or GFP. As the following chapter makes clear, fluorescence is a critical property of many modern stains, since it can "light up" structures of interest while leaving everything else dark. Fluorescence occurs when a molecule absorbs a photon of one color (the excitation wavelength) and emits a photon of a slightly different color due to its longer emission wavelength. Using filters similar in principle to (but a tad more sophisticated than) pink sunglasses, a microscope can allow the weaker emission wavelength through to the researcher's eyes, while blocking the much stronger (and messy) background illuminated by the excitation wavelength. Any structure tagged with such a molecule appears brightly lit, while the rest remain in darkness.

GFP is particularly useful because it is a genetically encoded fluorescent molecule, which thus can be switched on in any cell containing its gene, lighting it up from the inside. What is more, the gene for GFP can be cut and pasted immediately adjacent to the gene for any protein of interest, tacking on a sort of fluorescent spy that can track the protein under study as it goes about its business. It is difficult to overstate how profoundly GFP has revolutionized biomedical research, with applications in practically every one of its branches. On any given day, one researcher might be using GFP to track viruses in real time as they invade and infect cells, while another might be studying the development of synapses as they form in living tissue. Those who seek a touch of GFP magic in their living rooms may now purchase genetically modified (but otherwise entirely unharmed) glowing pet fish.

For applications in which several types of neurons or proteins must be studied simultaneously, there now exists a large range of differently colored GFP variants to distinguish them from one another. GFP's genetic basis has proved providential once again, as small modifications to the gene can cause the resulting protein to emit different colors. A few members in this large family of fluorescent proteins are shown in the image opposite, in which they have been purified and collected in vials. The top row shows them as they appear in white light; the bottom shows them fluorescing. GFP's colorful cousins were the beneficiaries of a bout of creative naming, and it is not uncommon to overhear heated debates in lab hallways over the relative virtues of, say, *mCherry* and *tdTomato*.

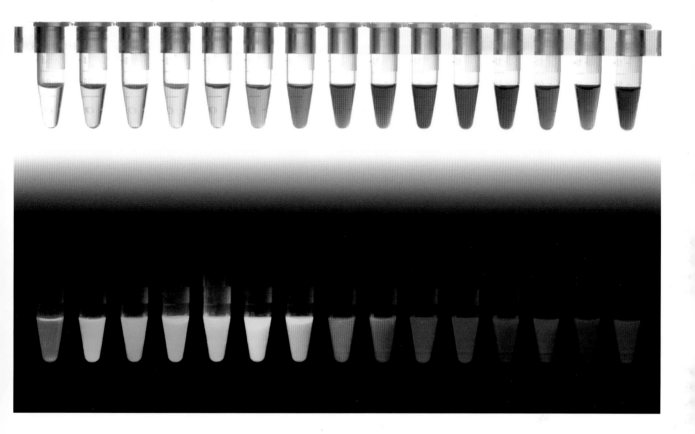

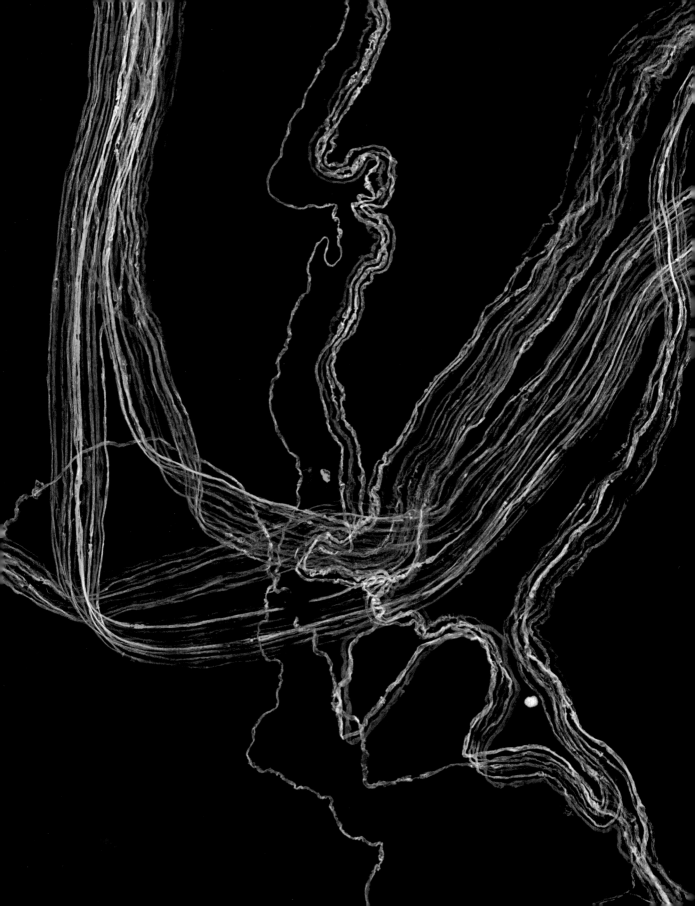

BRAINBOW

Opposite: Motor neuron axons.
Ryan Draft, Jeff Lichtman, and
Joshua Sanes, 2007.
Overleaf: Auditory brainstem.
Jean Livet, Jeff Lichtman, and
Joshua Sanes, 2006.

Camillo Golgi's *reazione nera*, described in Chapter 2, went a long way toward unearthing structures from the neural tangle. Yet his method and its modern heirs have a significant short-coming: They fail to unambiguously differentiate individual neurons illuminated in the same color. The problem arises from neurons' remarkable shapes: Unlike the large majority of cells in the body, which are small and roundish, neurons have dendritic and axonal extensions that travel long, tortuous paths and overlap significantly with their neighbors. For the neuroscientist in the business of working out how neurons connect to each other, the inability to distinguish one's appendages from another's essentially renders the task impossible.

An elegant piece of genetic trickery called Brainbow was recently proposed as a solution to this problem (see page 79). With the full power of modern genetics and molecular biology behind it, Brainbow is a logical extension of the research program initiated by Cajal more than one hundred years ago.

The pictures that have resulted from Brainbow, such as the one shown on this spread and the following three, are spectacular. Time will tell whether Brainbow can overcome the limitations of monochromatic staining strategies and deliver on its promise to uncover patterns in brain circuitry that still remain invisible to us. The image opposite shows several motor-neuron axons traveling side by side on their way to the muscles whose contraction they regulate. The image on the following spread zooms in on the *calyces of Held*, a complex knot of the largest synapses in the brain, located in the auditory brain stem.

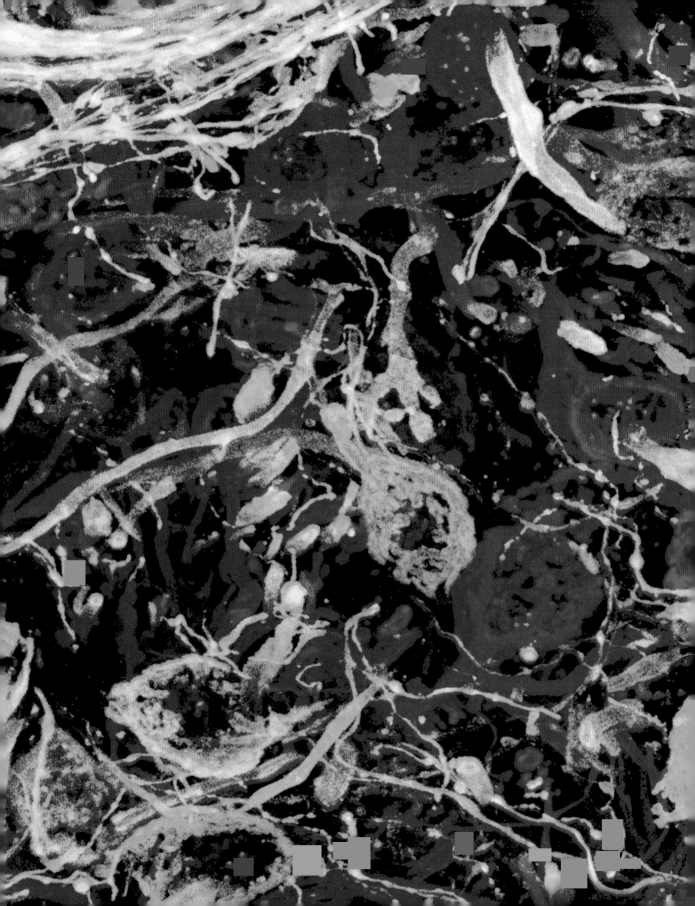

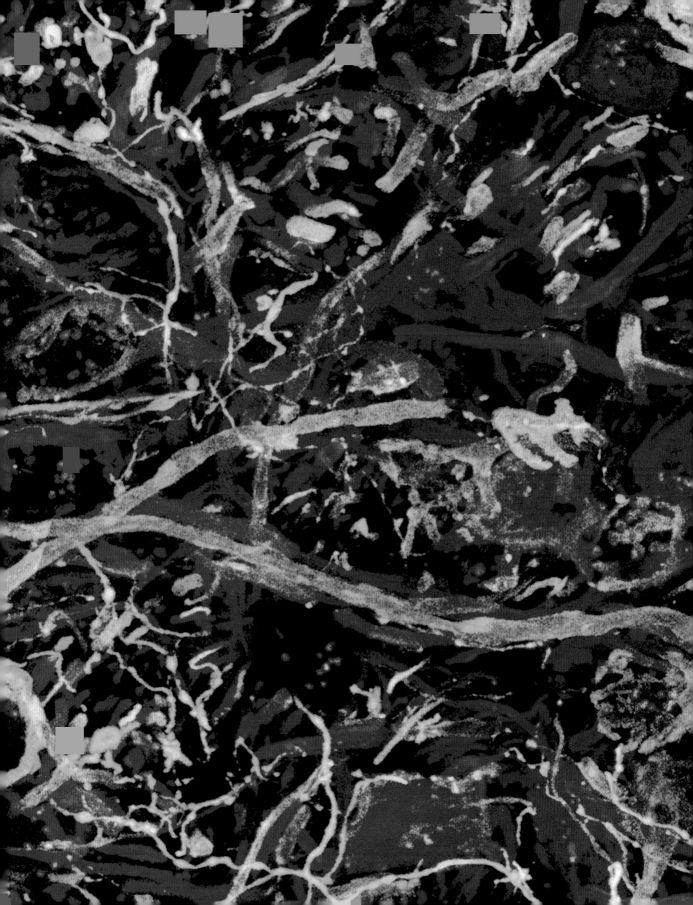

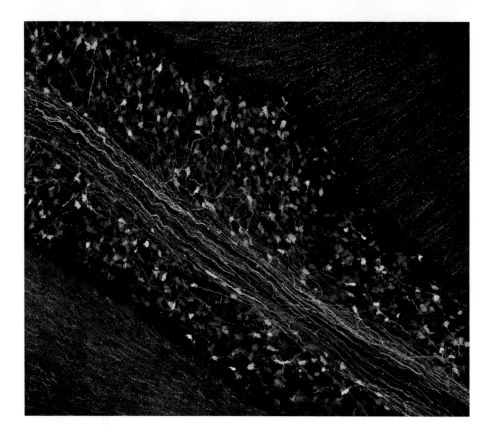

Above and opposite: Cerebellum.
Tamily Weissman, Jeff Lichtman, and
Joshua Sanes, 2007.

In the cerebellum (above), we see axons
coursing through the middle diagonal; the
colorful splotches on either side are enlarged
presynaptic terminals that the axons have
formed onto surrounding neurons. When he
saw them in Golgi-stained tissue, Cajal termed
these presynaptic terminals *rosettes* because
of their flowerlike appearance. The image
opposite shows a closeup of these rosettes.

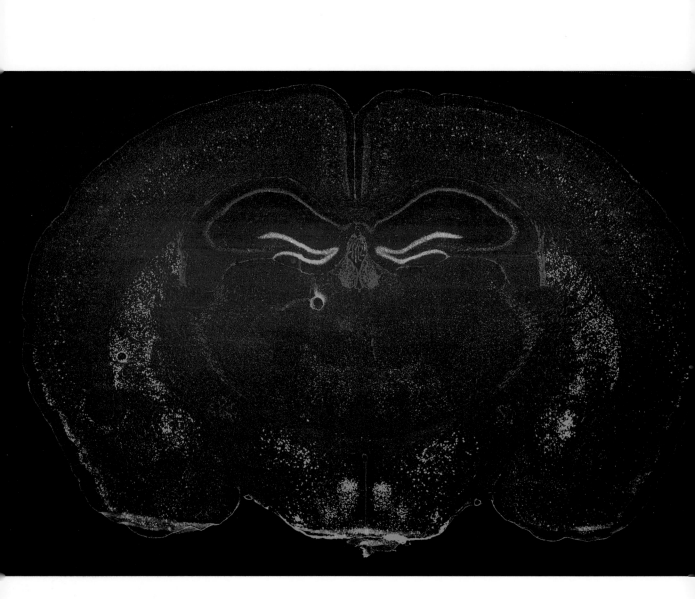

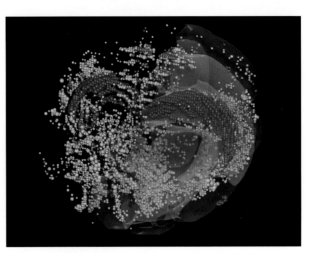

Opposite: In situ hybridization.
Allen Institute for Brain Science, 2006.

The data shown here were obtained using a method called *in situ hybridization*, which enables researchers to ask whether a given gene is switched on in a particular set of neurons. This knowledge is a critical step in developing genetically modified animals, such as the ones discussed on pages 82–87 and, more generally, in understanding what genes are switched on in different areas of the brain. In the example here, showing a thin slice of mouse brain, we find that two genes, one marked in red, the other in green, are switched on in populations of neurons that are clearly distinct from one another. The fact that these neurons are segregated suggests that they perform different duties, and that we can genetically tease one type of neuron apart from the other to uncover what these duties are. It is not so much that red- and green-marked genes are assumed to confer a special function on the neurons in which they are switched on; rather, these genes' specificity suggests that there is some deeper difference: The neurons they mark can be divided into two distinct classes.

Through painstaking in situ hybridization studies, researchers are gradually piecing together the patterns in which genes are switched on in the brain, deriving a taxonomy of neurons according to their genetic "signatures."

Above: Allen Mouse Brain Atlas.
Allen Institute for Brain Science, 2006.

The task of cataloging genomic maps of the brain, such as the image opposite, is far too vast for a single laboratory. Thus, as with the Human Genome Project, large consortiums and institutes have spearheaded the effort to sort out the details. Pioneering such enormous projects is the Allen Institute for Brain Science, which has generated data pinpointing where each of approximately twenty thousand different genes in the mouse brain is switched on, creating the publicly available Allen Mouse Brain Atlas. The image above shows the map throughout the entire brain for one of these genes, *Man1a*. The spheres mark areas in which *Man1a* is switched on, and their color adds an extra dimension by indicating how strongly this gene is "on" in each area, with red being the highest level. *Man1a* is striking for its high levels in the hippocampus, a region implicated in learning and memory. What significance this and many other gene maps hold remains unclear; this new field of study is still uncovering the borders of the countless territories delineated in the brain, and only beginning to explore the countries themselves. The Allen Mouse Brain Atlas was completed in 2006 and has since been employed in studies ranging from short-term memory to feeding behavior. One of the project's most striking results is the discovery that at least 80 percent of all genes in the genome are switched on in the brain, a stark reminder of its daunting complexity.

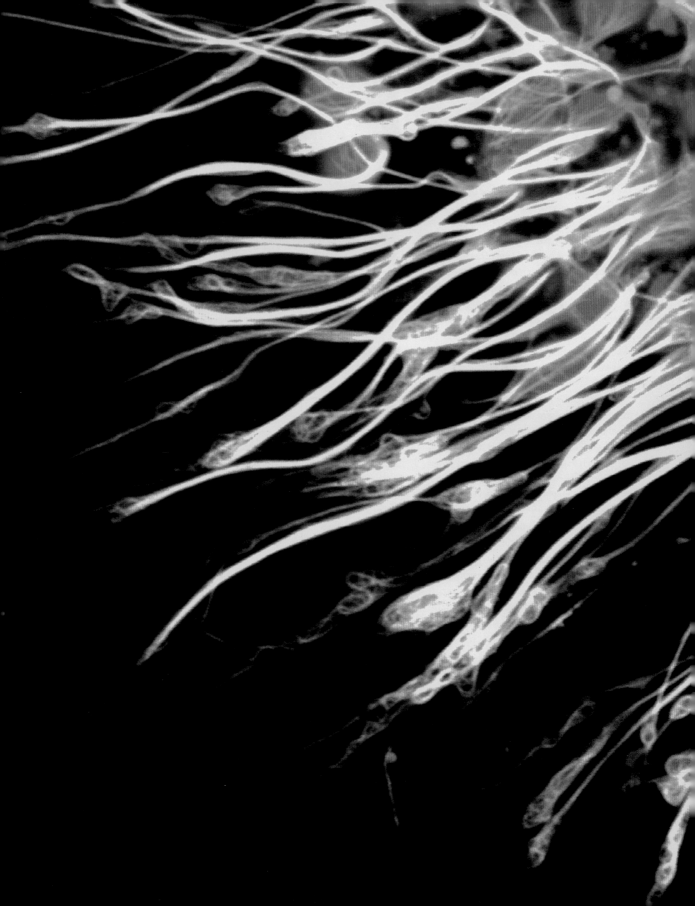

ANTIBODY STAINS

Opposite: Scaffolding in axons.
Michael Hendricks and Suresh
Jesuthasan, 2008.
Overleaf: Cerebellum.
Thomas Deerinck and Mark Ellisman, 2004.

In situ hybridization is a critical tool for determining which genes are switched on in a given
cell type. However, the method cannot resolve
where within a neuron specific proteins can be
found, an often critical clue about what duties
they perform. A protein found only in synapses,
for instance, is likely to be involved in the
transmission of information from one neuron to
another. Conversely, if a protein's job is already
well known, its position in neurons or whole
tissues can suggest how they perform their
functions and offer insight into how they
malfunction in cases of disease. To address
those questions, neuroscientists have adopted a
now ubiquitous method in biochemical research
called *immunohistochemistry*. Like in situ
hybridization, it takes advantage of a basic
biological mechanism, in this case via antibodies,
the henchmen of the immune response.
Antibodies are so effective because they can
recognize, and strongly attach themselves to,
molecules introduced from outside an organism's
body, such as those lining the surface of
pathogens. Biologists have developed ways to
harness the powerful ability of antibodies to
recognize specific molecules and can employ
them to study any protein of interest in the brain.
By revealing where a given protein is found in a
tissue and even within an individual cell,
scientists are afforded precious insight into a rich
molecular world otherwise invisible even under
the microscope.

In the image opposite, an antibody that
stains the scaffolding of axons growing in a dish
reveals their shape (yellow). The next four
images, all of which are pseudocolored, are
examples of different tissues stained using
immunohistochemistry. The image on the
following spread shows the cerebellum in which
two cell types—glia and Purkinje neurons—can
be distinguished: Red is an antibody staining of
GFAP, a cytoskeletal protein found in glia, the
non-neuronal cells in the brain that protect and
support neurons with nutrients and oxygen.
(Since GFAP is found almost exclusively in glia,
its presence can be used to distinguish them
from neurons.) Green is an antibody staining
of IP3 receptors, proteins that release cellular
calcium stores. Purkinje cells—like the one
drawn by Cajal on page 56—are rich in IP3
receptors, so this antibody staining reveals
these neurons' characteristic shape in the tissue.
The structures in blue are stained with Hoechst
33342, which is not an antibody but a small
chemical that binds strongly to DNA and so is
often used to stain cell nuclei.

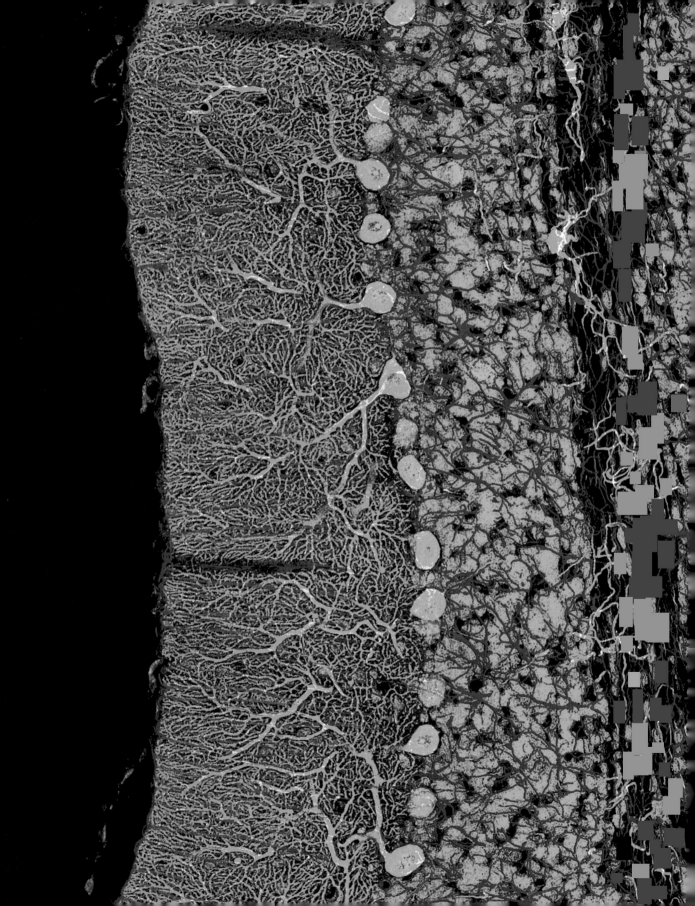

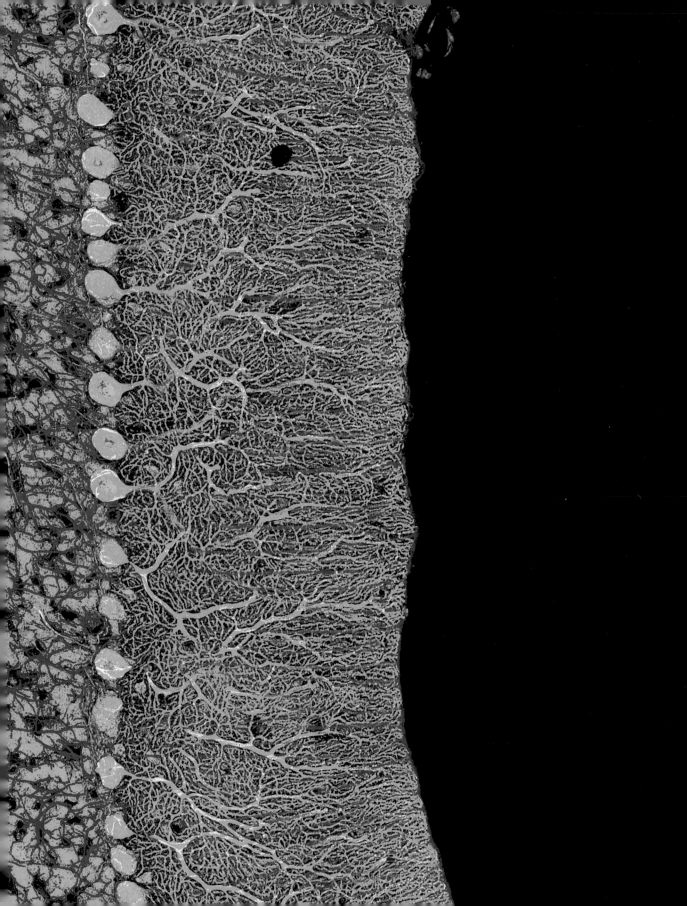

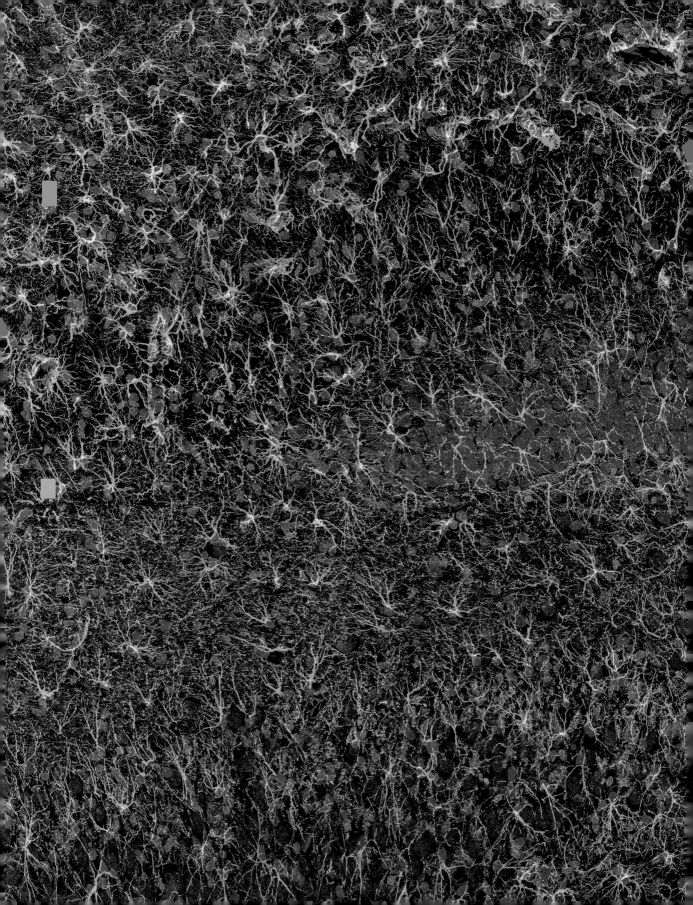

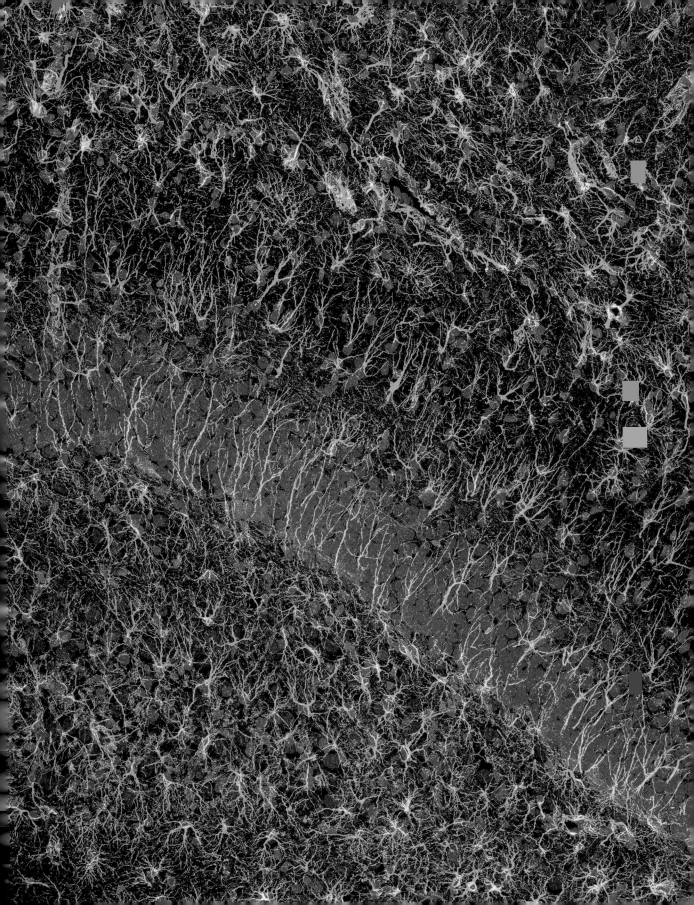

Previous spread: Hippocampus.
Thomas Deerinck and Mark Ellisman, 2004.
Opposite: Glial cells.
Thomas Deerinck and Mark Ellisman, 2008.
Overleaf: Serotonin in the cerebral cortex.
Lasani Wijetunge and Peter Kind, 2008.

The image of the hippocampus on the previous
spread was obtained with an antibody staining
of GFAP (now in green), along with an antibody
staining of Neurofilament 68kD (blue), another
cytoskeletal protein specific to neurons and found
mainly in their axons. Propidium iodide (red), like
Hoechst 33342, binds DNA and is used to stain
cell nuclei.

In the image shown opposite, glia are
illuminated in the cerebellum, once again with
an antibody staining of GFAP (yellow). The image
on the following spread reveals the location of
a transporter of serotonin (a widespread neurotransmitter) in areas of the cerebral cortex
that process sensory information such as touch
and hearing.

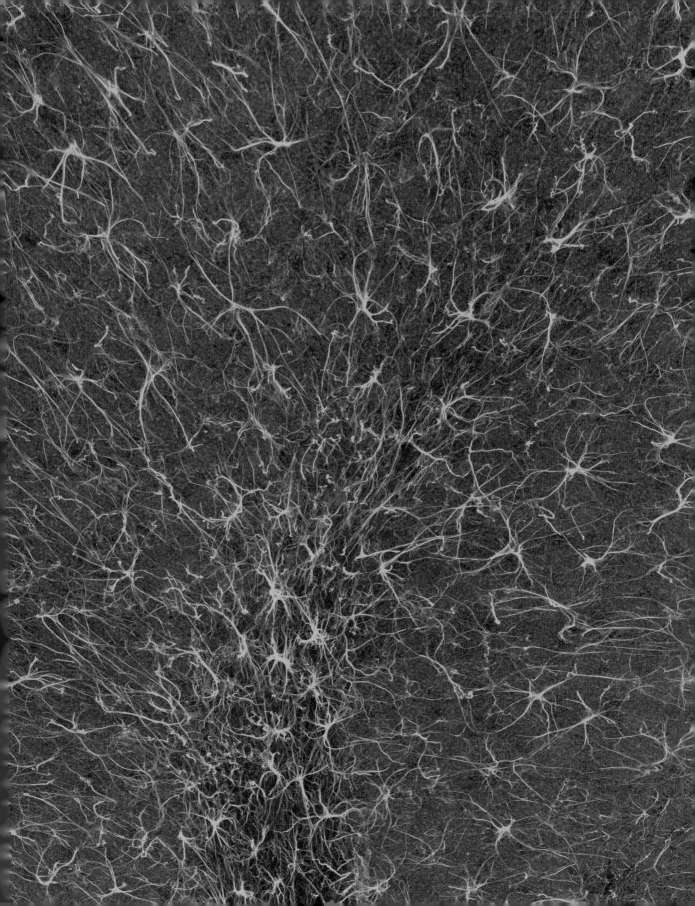

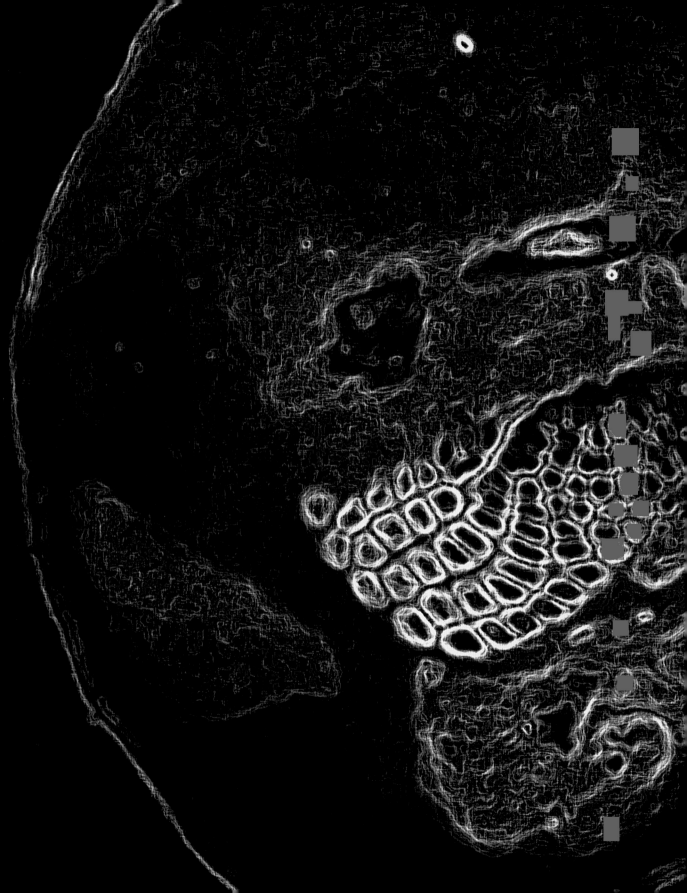

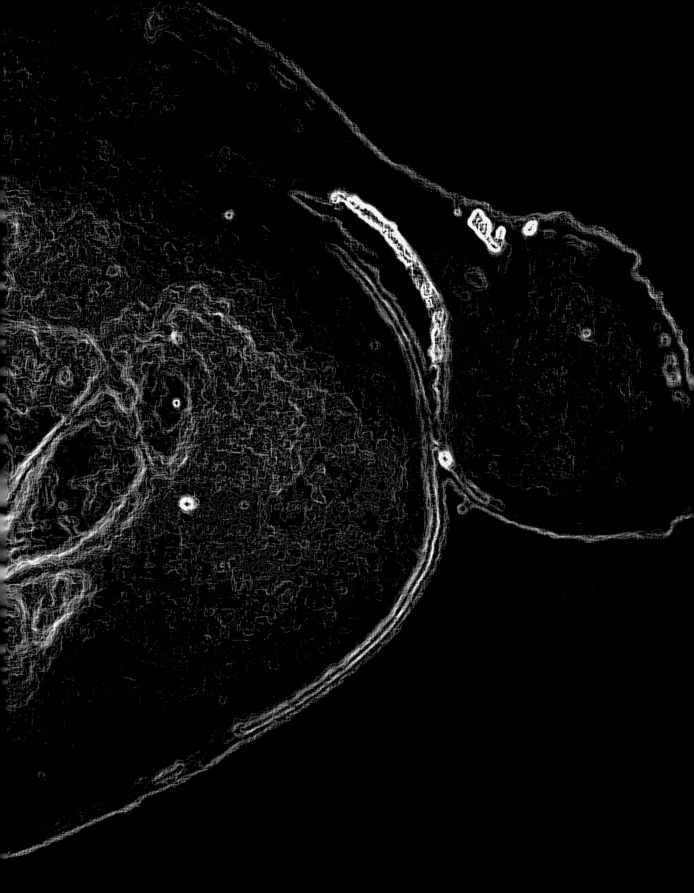

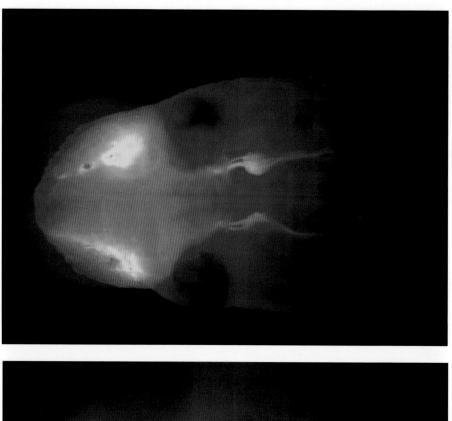

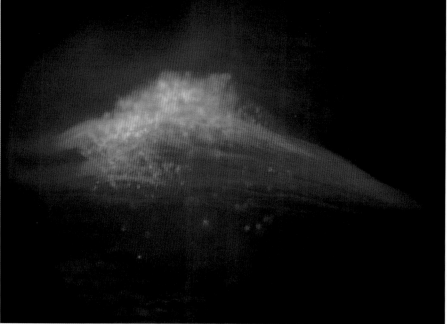

PORTRAITS OF THE MIND

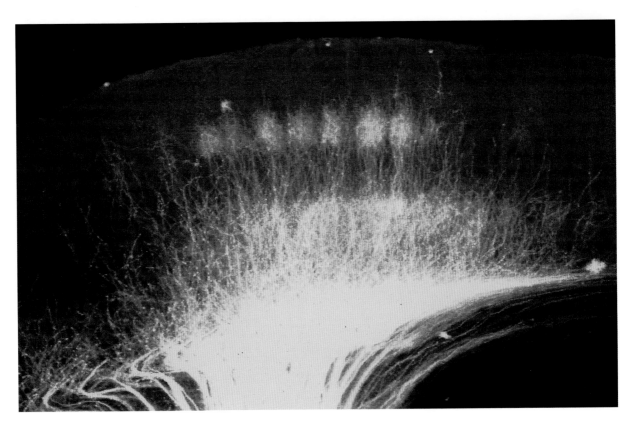

DiI

Opposite: Pathway from whisker to brain.
Lasani Wijetunge, Alex Crocker-Buque, and
Peter Kind, 2009.
Above: Barrel cortex.
Aric Agmon, 1992.

These three images illustrate techniques
employed to understand how neurons in the brain
are wired up to one another. *DiI* (pronounced
dye-eye) is an oily, nonbiological, fluorescent
chemical that infiltrates the fatty membranes of
axons and dendrites and, over a period of hours,
days, or even weeks can inch its way down their
long spans across the brain. In this way, DiI
exposes which areas of the brain the axons of a
particular neuronal population travel to, providing
a snapshot of which ones are communicating.

In the images opposite, DiI (shown in
bright orange) was delivered to neurons that
sense the vibrations of whiskers on either side of
a mouse's snout (the entire head is pictured in
the top image, with the snout pointing left). These
neurons' axons, now illuminated in bright orange,

can be seen traveling from the snout toward
the brain (not visible). The image below is a
closeup of the somata in a bundle of neurons
carrying the information.

In a separate experiment (above), DiI
was delivered to neurons in the thalamus, a
relay in the pathway from whisker to brain.
The projections from the thalamus terminate in
the middle layer of the neocortex, where all of
the axons conveying information from a given
whisker form a specific cluster (appearing
toward the top of this image and taking on
a rectangular appearance). Each of these
clusters is enclosed within a "shell" made up
of cell bodies (not labeled), dubbed a "barrel."
(In the first detailed account of barrel anatomy,
published in 1970, the authors were so struck—
and presumably amused—by their uncanny
resemblance to wine barrels that they included
a seventeenth-century etching of one, for
comparison.) Each one of these fascinating
barrel structures corresponds to a single
whisker and is dedicated to processing the
tactile information it obtains from the world.

Diolistics.
Selva Baltan and Jaime Grutzendler, 1999.

This image was generated using the *multicolor diolistic* method, in which tiny metal particles are coated with differently colored variants of DiI and then scattered into the brain tissue with a pressured-air "gun." DiI outlines white-matter structures (axons and glia) at random, which can then be distinguished from one another by their varied coloration. In many cases, the entire architecture of white-matter structures can be lit up and visualized in this way. This image was taken in the *corpus callosum*, a prominent white-matter structure. Here we do not see entire axons or glia, but just segments traveling through this particular slice of brain.

Biotinylated dextran amine.
Nathaniel Sawtell, 2009.

This image shows a small molecule called
biotinylated dextran amine (BDA) delivered
via injection (the dark spot at top right) into
a fish brain and taken into the extensions
(dendrites and axons) of neurons that happened
to be traveling through the neighborhood. Once
inside a neuron, BDA spreads across its entire
span, revealing dendrites (the vertical lines) and
their attendant round somata (at bottom right).
Axons can also be resolved (shown sweeping in
from the top left), presumably forming synapses
onto the dendrites. Since BDA fills the entire cell,
it is possible to locate, sometimes as far as
millimeters away, somata whose axons were
present at the point of BDA injection and so piece
together the neuronal circuit.

Rabies.
David Lyon and Edward Callaway, 2005.

The late Francis Crick, who turned his attention to neuroscience after his seminal work on the structure of DNA, once wrote that the field needs "a technique for injecting a single neuron in such a way that all the neurons connected to it (and only those) are labeled."[1] Would such a technique unearth patterns of connectivity or a network that's random and unstructured? Some hope that this knowledge, currently beyond our reach and buried deep within the tangle of neurons in the brain, will allow us to make inferences about function from the underlying anatomy.

Although still brand-new and actively being developed at the time this book went to press, an elegant method has been devised to make good on Crick's request. Like many of the powerful techniques described in this chapter, it borrows a natural process and retools it. In this case, nature's contribution is the rabies virus, which derives a devastating potency from its ability to work its way up a neuron's axon, into its soma, and on to its dendrites, then cross synapses into the axons of the connected cells.

The virus propagates against the direction of information flow in the neuron; by infecting motor neurons in, for instance, an arm bitten by a rabid dog, it crosses synapses "backward" all the way up to the brain, where it spreads quickly because of the neurons' high degree of interconnectivity, rapidly killing the host.

From this grim process, a silver lining: Researchers have recently engineered the gene for GFP into the rabies virus. In addition, they pseudotyped the virus and deleted a critical gene from the rabies genome that allows the virus to spread, so that only those neurons directly connected to the original host neuron become infected. This image reveals a tantalizing result: A researcher has injected the neuron shown in red with the modified rabies virus, revealing in yellow the neurons directly connected to it.

This journey from dog bite to neuroscience lab is just a variation on a persistent theme: Our most powerful means of uncovering nature's secrets is not always to design agents from scratch. Instead, research has often thrived by adapting nature's own mechanisms, having evolved over eons, to our current needs.

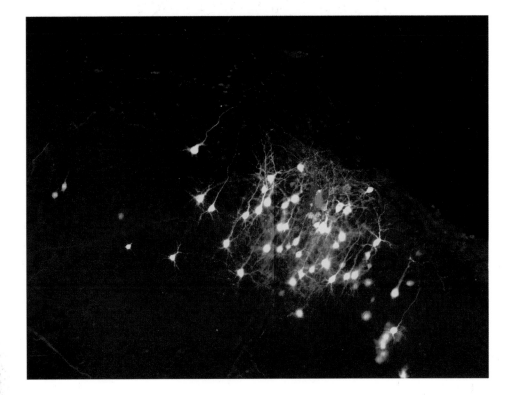

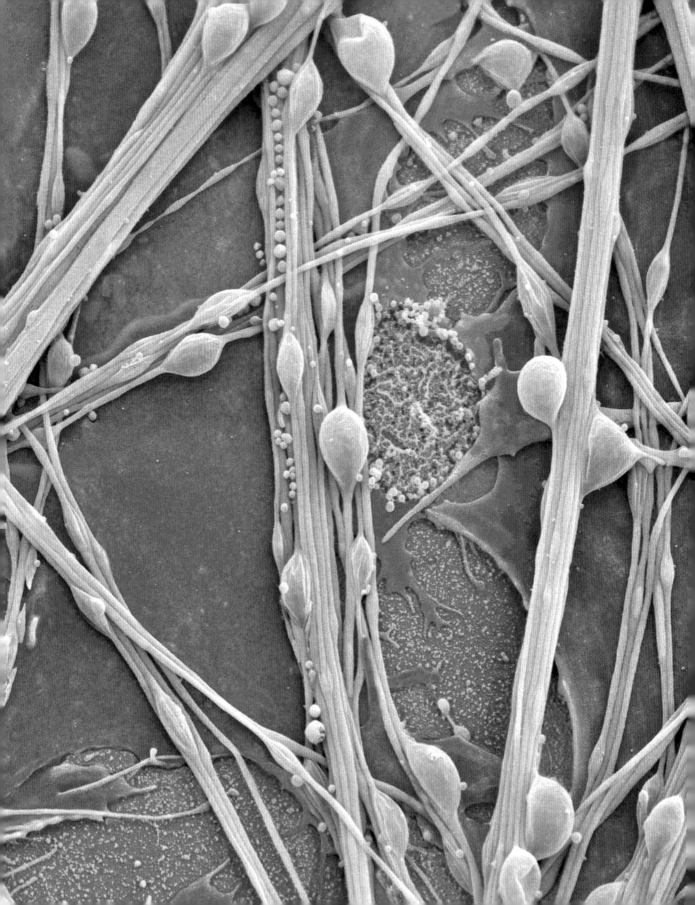

4.

BREAKING THE DIFFRACTION BARRIER
FROM CELLS TO MOLECULES

BREAKING THE DIFFRACTION BARRIER:
FROM CELLS TO MOLECULES

by Maryann E. Martone and Mark H. Ellisman
National Center for Microscopy and Imaging Research
University of California, San Diego

The nervous system is no doubt the preeminent challenge to those pioneering advanced imaging technologies. The most complex structure in the known universe, it weaves together miles of microscopic cellular cables that carry signals to trillions of tiny synaptic contacts. An even finer meshwork of thread-like filaments extends *within* each one of these cables, endowing neurons with their shape, providing mechanical stability to the brain, and functioning as molecular highways between different parts of the cell. Sometimes these nanoscopic filaments form an element of the synaptic connections between two neurons, serving as a transcellular scaffolding that facilitates tight interactions that occur between certain nerve cells. Moreover, the entire organ is constantly changing over time, on time scales from milliseconds to decades. Depending on your perspective, this is either a microscopist's dream—or her nightmare.

Hidden somewhere in the tangled mass of the brain lies the biological basis of thinking, acting, remembering, feeling, forgetting, and desiring. Occasionally something goes wrong, leading to system breakdowns like Alzheimer's disease, schizophrenia, or drug addiction. These cognitive and pathological states could arise from small molecular imbalances that are replicated across thousands of cells; or they could involve changes in the sheer numbers or structures of neurons; alternatively they could be caused by subtle alterations in the way that a few neurons transmit information in a very localized region. Whatever the mechanism, we must somehow develop the means to detect and measure these factors in order to understand how the brain makes us who we are, and what happens when things go wrong.

The ultimate microscope would allow us to zoom in continuously from a view of the entire organism, such as a human or mouse, down to the individual molecules that constitute its cells and the tissues they form. It would present this data in three dimensions, revealing the precise way in which neurons are wired up to one another, and it would include the ability to navigate a critical fourth dimension—time. This would equip us to observe dynamic behaviors unfolding over periods ranging from microseconds to weeks or months. This God's-eye view would enable us to explore everything there is to know about the structures that govern processes like cell division, tissue repair, and the mechanisms used by synapses to change over time.

Unfortunately, no such ultimate microscope exists as of yet, so we have had to get clever with what we've got. We already have at our disposal a battery of marvelous instruments and techniques that are capable of revealing truly astonishing features. Each one has its particular strengths but none can achieve everything on its own. Our ability to detect small structures is still hobbled by certain fundamental problems, like the limits set by the physics of light waves (see page 123), or the harsh treatments required for certain specimen preparations. Moreover, the fluorescent stains that have proven so powerful and versatile in light microscopy, notably for studies in live tissue, are exceedingly difficult to adapt to electron microscopic imaging. And although the electron microscope can detect much smaller structures than those visible with fluorescence, its application requires placing the specimen in a high-vacuum, high-radiation environment, thus completely ruling out studies in live brains.

Because of these substantial limitations, we are faced with important resolution gaps in our view of certain types of structures. For instance, dendritic spines are too large to be imaged easily in electron microscopy without some sort of three-dimensional reconstruction, yet possess features too fine to be discerned in sufficient detail with a light microscope, even when applying the newest super-resolution methods. We know that dendritic spines are dynamic and can change their size, shape, and number as a function of experiences like acquiring new memories, so the stakes for understanding them in detail are extremely high. Yet until the ultimate microscope arrives, we will be forced, like the blind men and the elephant, to draw imperfect conclusions from information gleaned on both sides of the gap.

In very recent years we have witnessed a convergence of technological advances, some truly revolutionary, that promise to help bridge the resolution gap. First, new microscopy methods that are being developed offer marked improvements in resolution and versatility. For instance, a brand-new electron microscopy strategy (see pages 131–135) is capable of providing images at standard electron microscopic resolutions but across three-dimensional spatial scales never before imagined. Second, as microprocessors become cheaper and faster our computational power continues to increase, enabling sophisticated image analysis that can sometimes overcome traditional resolution limits. And third, thanks to the molecular revolution, we are beginning to develop stains that can reveal structures in living systems using light microscopy, which can then be adapted to electron microscopy applications in the very same sample.

While the insights into brain structure produced by these microscopy techniques are stunning, we may still be tempted to ask ourselves if that is it. Is there something missing, some "dark matter" of the brain, perhaps, that our staining techniques have failed so far to reveal? If you examine electron micrographs very carefully, your eye might pick up on the pale ghosts of tiny filaments that appear to connect across cell membranes. These transcellular filamentous networks could be forming the scaffolding for a yet unknown kind of signaling that operates independently of classical synaptic transmission. Are these specters real or mere artifacts? We are certain of nothing yet, and a conclusive answer requires further progress in the explosion of novel strategies that have marked the past few years. We inch forward into the darkness with only the reassuring knowledge that entire branches of neuroscience have sprung out of innovations in imaging, and the hope that this pattern will continue to hold far into the future.

Confocal microscopy: dendrite.

Carl Schoonover and Randy Bruno, 2009.

All the sophisticated staining methods described in Chapter 3 would be useless without the right microscope to detect their patterns. Novel microscopy methods played a significant role in revolutionizing neuroscience in the nineteenth century by enabling scientists to observe previously inaccessible structures. Likewise, innovation throughout the twentieth century has fueled discovery by allowing us to resolve the fine details uncovered by our growing tool kit of fluorescent markers. A relatively recent addition to the neuroscientist's arsenal is the confocal microscope.

Traditional microscopes cannot observe a structure inside a biological sample without also detecting out-of-focus objects above and below it. This results in a confusing, blurry picture—especially when the sample is as tangled as brain tissue. A confocal microscope elegantly solves this problem by collecting light only from its lens's plane of focus, ignoring structures on either side of it. A laser beam scans across the plane to illuminate fluorescent molecules in its path, creating a series of artificial optical "slices" through the thickness of an entire specimen. The microscope then collects light from the structures that are in focus in each optical slice. This image shows thirty-two such slices (small panels) obtained in a single sample of a rat's neocortex. The top-left slice was captured at the surface of the sample; each successive slice (arranged in columns) was obtained by minutely shifting the specimen closer to the lens, in order to image a bit deeper, creating a stack of thirty-two optical slices through the sample. The large image at the bottom was created by summing up all of the slices, revealing a dendrite and its spines.

Because of their marked improvement in clarity, confocal microscopes have become standard across the field of neuroscience, and indeed biomedical research at large. Many of the images in Chapter 3 were acquired using this tool.

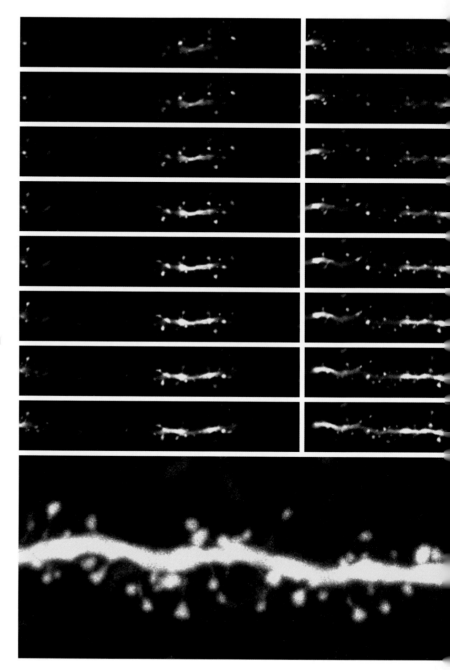

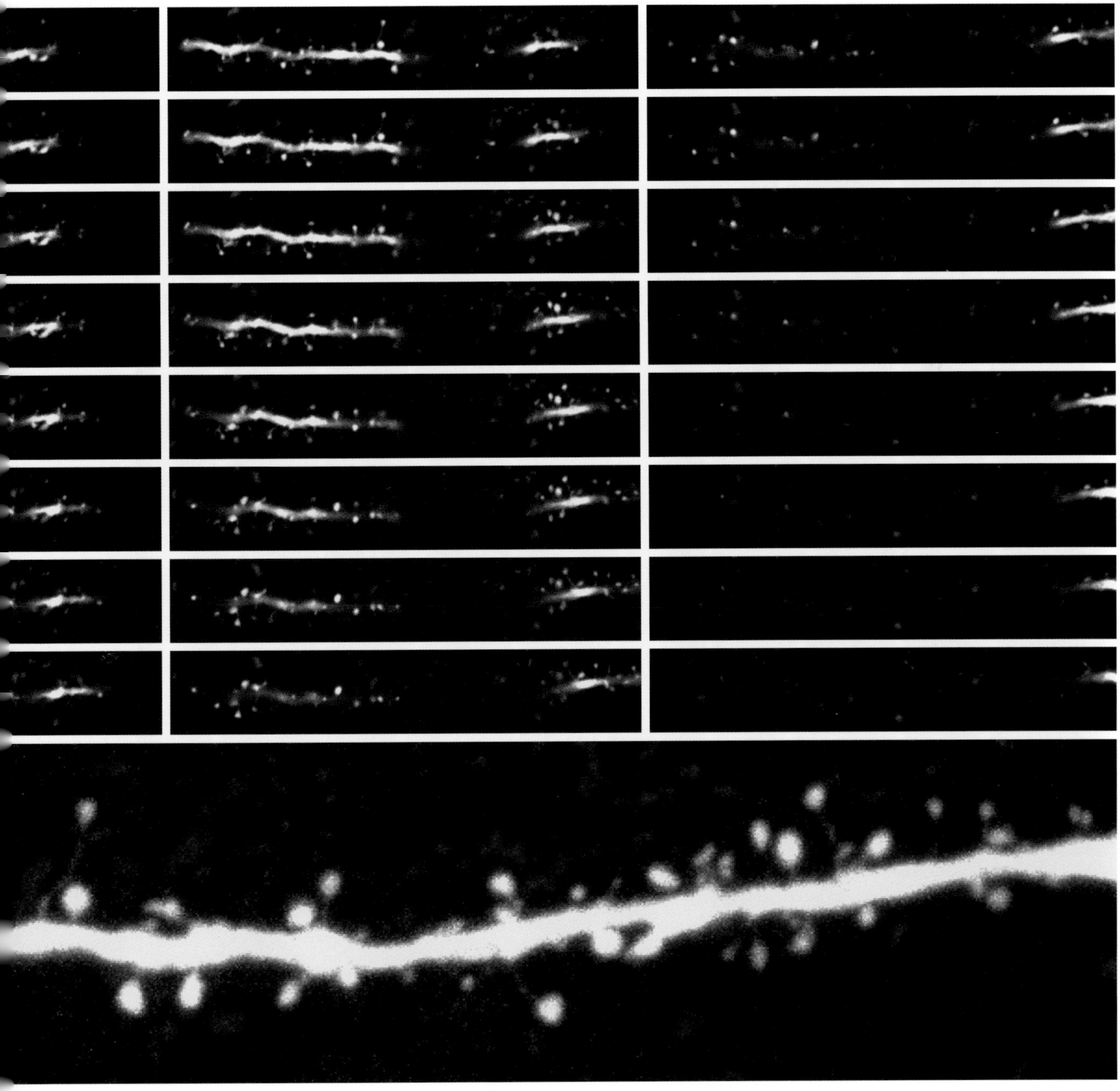

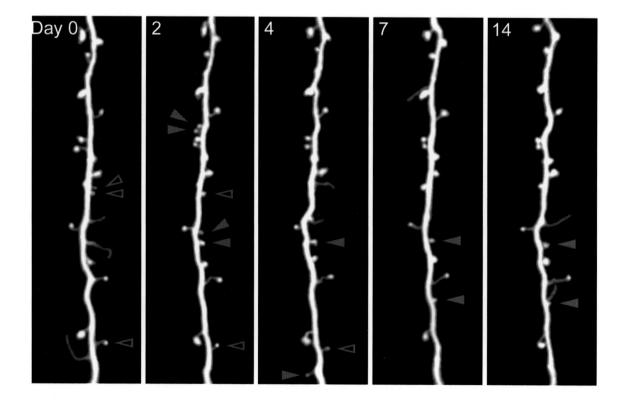

PORTRAITS OF THE MIND

**Two-photon microscopy: spine dynamics
in a living animal.**
Guang Yang and Wenbiao Gan, 2009.

Confocal microscopy works beautifully for thin samples, such as a fine slice of brain on a slide. But because light beams are scattered by biological tissue, this method cannot be used to image structures deep within a sample, such as neurons inside a live brain. The two-photon microscope solves this problem by exploiting a quirky phenomenon of physics, first formulated in the 1930s. It states that two particles of light, or photons, will behave as if they were a single one (with double the energy) if they hit a fluorescent molecule almost simultaneously. In the context of two-photon microscopy, the odds of this happening are exceedingly slim except at a microscope's focal point. This phenomenon enables two-photon microscopes to illuminate only the structures that fall within that tiny point, conveniently ignoring the rest. Compared to its confocal cousin, this class of microscope can access structures deep inside a sample, while causing less damage to the tissue. For these reasons it has become the tool of choice for studies in living brain tissue. This advantage stems from that 1930s finding, which ensures that light-scattering tissue between the lens and the region in its focal point is effectively rendered invisible since a pair of photons will practically never meet on their journey through the brain until they hit that spot. Moreover,

two-photon microscopy employs infrared light, which can travel across biological tissue much more easily than visible light (it also does less damage to the tissue). The tool's penetration power is such that neurons in a live brain can be imaged simply through the animal's thinned-down skull.

In this example, two-photon microscopy was used to track the same dendrite in a living mouse brain over a period of fourteen days, revealing stunning dynamics of spines appearing and disappearing. Here was a direct visualization of plasticity occurring in the brain. The study from which this image was taken found that the formation of new spines—and so, presumably, new synapses—correlates with novel experiences. Its authors deduced that some of these new spines are preserved for an entire lifetime, a phenomenon that would enable the animal to maintain lifelong memories. The dendrite imaged in the left panel (Day 0) was obtained before exposing the animal to a highly stimulating environment it had never previously experienced. On Days 2, 4, 7, and 14 after this exposure, the exact same area of dendrite was reimaged, revealing newly formed spines (filled red arrowheads) and eliminated ones (open arrowheads). This astonishing front-row view of the plasticity of connections between neurons suggests how their dynamics may enable the brain to acquire and maintain memories.

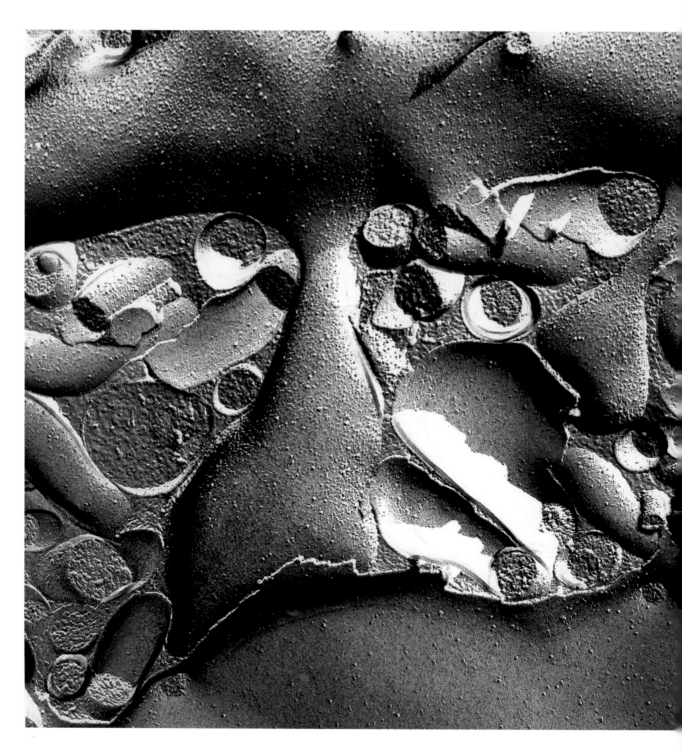

PORTRAITS OF THE MIND

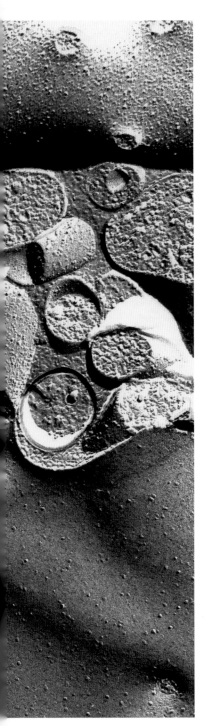

Electron microscopy: dendrite.
Kristen M. Harris, 1986.

One of the greatest difficulties of neuroscience is that many of the objects most relevant to the brain's function are too small to see clearly, even with the most sophisticated light-based microscopes. The "size" of visible light—the distance between each peak of light's wavelike trajectory through space—measures about half a micrometer. A minute measurement, to be sure, and yet many of the interesting molecules and structures in the brain are orders of magnitude smaller. Trying to "see" them using light is akin to typing with boxing gloves: The probe is unfit for the object.

This problem has been known for some time, with its formulation dating back to the nineteenth century. The diffraction barrier is a law of physics, and no amount of technical brilliance or sophisticated gear will make it go away. But there are ways to sneak around it, and scientists have developed methods for resolving structures—such as the inside of a synapse—that are far too small to be visualized under conventional light-based microscopy.

One solution, already decades old, is to abandon light in favor of electron beams. Like photons, electrons travel in a wavelike fashion, but crucially, their wavelength is far smaller than that of light, making their resolving power far greater. This image, obtained using an electron microscope, shows the spines on a dendrite at a resolution unattainable by even the most sophisticated light microscope. Running across the top of the image is the cylindrical dendrite, from which two spines can be seen budding off and pointing downward—compare these to the spines imaged using light-based microscopy on pages 118–120.

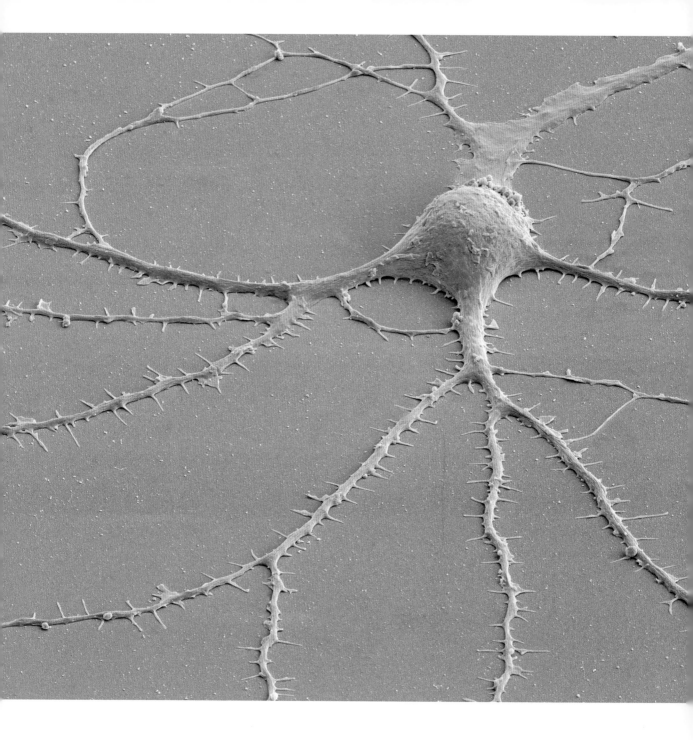

PORTRAITS OF THE MIND

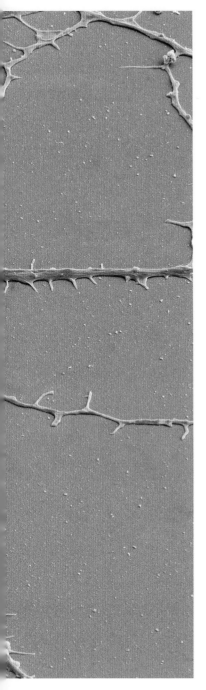

SCANNING ELECTRON MICROSCOPY

Opposite: Spiny neuron.
Below: Axons and boutons.
Thomas Deerinck and Mark Ellisman, 2009.

Electron-based microscopy comes in two primary flavors: scanning and transmission. The next three images focus on scanning electron microscopy, in which a beam of electrons is scanned across the surface of a sample, and a detector keeps track of electrons bouncing off its surface, revealing the specimen's outer shape. The image opposite shows a soma with dendrites radiating from it; their spines stick out much like in Cajal's spine illustrations (see page 59). The image below shows axons visualized in this manner. Note their teardrop-shape boutons swelling off the main shaft. (Pseudocoloring helps to differentiate elements in the image: The brown background is a surface of supportive glial cells, upon which the beige-colored neurons were grown before imaging; the treated glass surface they were cultured on appears in green beneath them.)

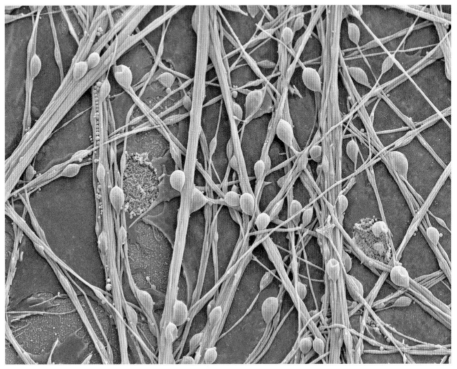

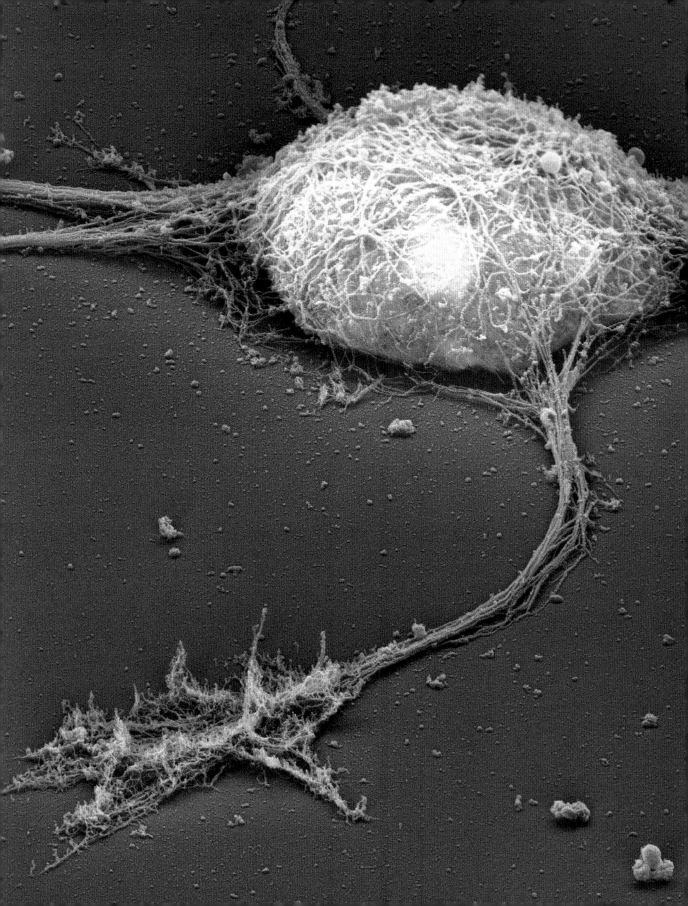

Opposite: Growth cone.
Bernd Knöll and Jürgen Berger, 2008.

The image on the opposite page undoubtedly would have pleased Cajal, who devoted years of painstaking study to the growth cone—the motile tip of a young neuron's burgeoning axon, which extends out in search of partners with which to form synapses. It is seen here at the bottom left, with the soma and other extensions at the top right. Cajal compared his beloved structure to a "living battering ram, soft and flexible, which advances, pushing aside mechanically the obstacles which it finds in its way, until it reaches the region of its peripheral distribution."[1]

Since the electron microscope did not exist in his era, Cajal was unable to make out the dense internal meshwork of long, solid cables that form the growth cone's skeleton. Long before we developed nanowires, nature came up with a way to endow soft, amorphous cells with solidity and structure. To reveal the growth cone's insides in this image, the outer plasma membrane of the cell was removed and the remaining structure, the nanowire scaffold, was dried and coated with metal.

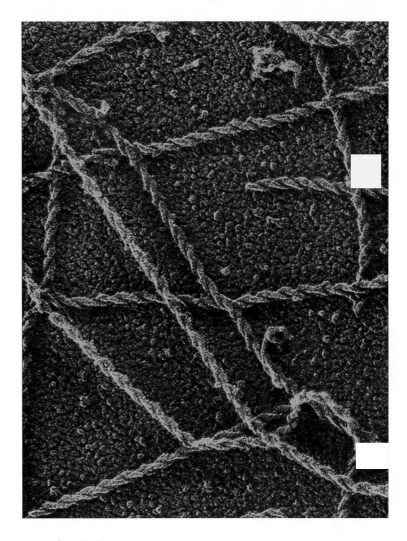

Above: Cytoskeleton.
John Heuser, 1985.

A closeup of *actin filaments*, the strong fibers that form an important part of the cell's skeleton, reveals their exquisite organization. This image was obtained by purifying these filaments, which are composed of countless small, interlocking actin proteins. In order to accentuate the filaments' helical pattern, a *myosin head*, the purified elements of another protein that attaches itself strongly to actin, was introduced into the preparation, bestowing their thick, ropelike appearance under the electron microscope.

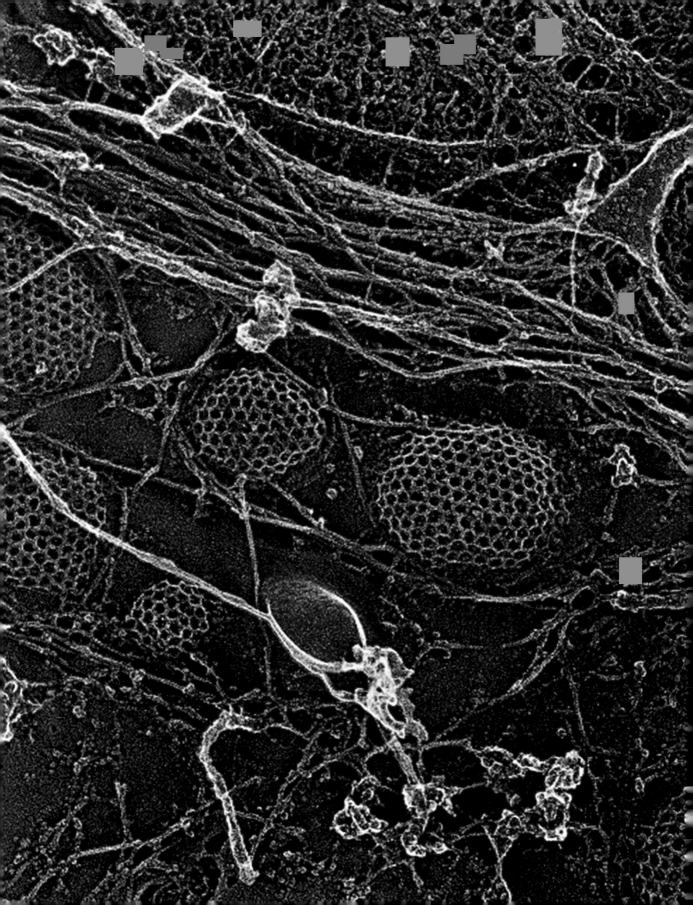

Vesicle formation.
John Heuser, 1982.

Neurons communicate with each other at the synapse by releasing neurotransmitters such as dopamine and serotonin in a process called synaptic transmission. In order for synaptic transmission to occur quickly and reliably, every bouton mobilizes an intricate arsenal of molecular devices. In brief, an electrical impulse travels down the axon to a neuron's boutons, which contain dozens of balloon-shape vesicles laden with neurotransmitter. These vesicles dump their contents into the synapse, activating receptors in the neurons downstream. The maintenance and regulation of neurotransmitter vesicles is a fundamental aspect of neuronal communication; their malfunction is linked with devastating conditions such as epilepsy and schizophrenia. Unsurprisingly, some agents that target the nervous system act on parts of this machinery. The one commonly known as Botox, for instance, doesn't paralyze the muscles themselves, but instead disrupts the neurons that regulate their contraction by immobilizing their vesicles and preventing them from delivering neuro-transmitter. (Since it disrupts such a core process, Botox is exceptionally lethal. Properly administered, five hundred grams would be sufficient to wipe out half the human race.)

This image, although not from a neuron, shows only one part of the process underlying the proper regulation of vesicles: the mechanism through which they are created out of the cell's plentiful membrane. Each geodesic structure captured in this image, obtained using an electron microscope, is one such vesicle in the making. They are covered with an exquisite molecular lattice that forms into a sphere and ultimately, improbably, tears the newly minted vesicle out of the cell's membrane.

This specimen was prepared by blasting the inside of a fibroblast cell with loud, high-pitched sounds. Then, to preserve its structure, the sample was rapidly cooled in a device called a "freeze slammer," which rams it into a slab of copper just four degrees above absolute zero (-273 degrees Celsius).

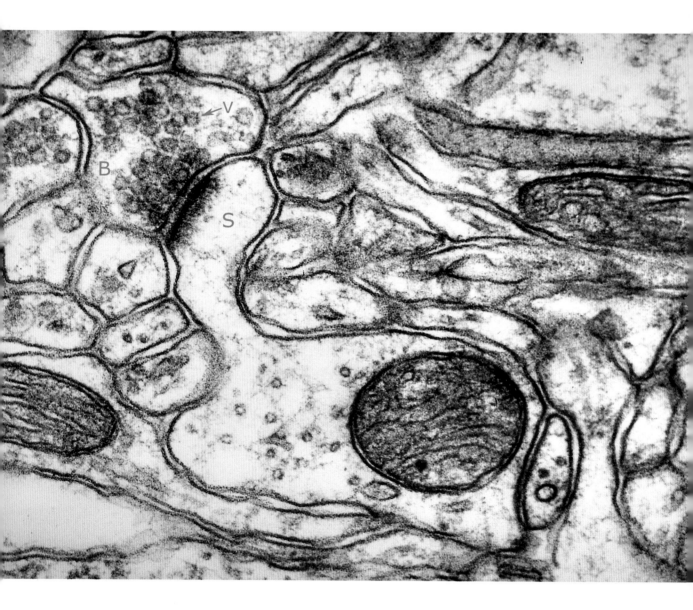

PORTRAITS OF THE MIND

Opposite: Synapse.

Josef Spacek and Kristen M. Harris, 2000.

The second major electron-based microscopy method, transmission electron microscopy (TEM), affords a view through biological samples, rather than detecting their outer shape. It functions by shining a beam of electrons through an infinitesimally thin slice of the specimen. The parts of the sample that absorb electrons (such as dense clusters of proteins) turn up dark in the image that is detected on the other side of the beam source; in areas where the beam crosses the sample unobstructed, the image is brighter. The image opposite shows the synapse under this different perspective. (B labels a bouton, S the spine it is connected to.) The spherical vesicles are sliced through the middle and appear as circles (V). (Note that although they appear as having been taken with a scanning electron microscope, the images on pages 127 and 128 were actually obtained using a transmission electron microscope.)

SERIAL BLOCK-FACE SCANNING ELECTRON MICROSCOPY

Overleaf: Individual sections.
Page 134: Five cells reconstructed by computer from eight hundred sections.

Thomas Deerinck and Mark Ellisman, 2009.

So why not scrap light microscopy entirely and rely fully on the superior illuminating power of electrons? For one, electron microscopy is exceedingly time-consuming, which restricts its scope to tiny areas. It requires great patience and not a little technical brio. An electron microscopist must painstakingly cut hundreds of consecutive ultrathin slices of a specimen, then capture an image of each one on the microscope, and finally align all the images to reconstruct a three-dimensional representation of the intricate structures from the slices. Scalability is a major, sometimes fatal issue, and it is impossible to study large areas of tissue in this way.

A recent innovation may lay this problem to rest, at least for some applications. Serial block-face scanning electron microscopy promises to automate, streamline, and markedly accelerate the procedure described above. It works by placing a thick sample into a scanning electron microscope, shining an electron beam onto it, and collecting the electrons that bounce off. (Although technically a scanning electron microscope, it approximates the view of a transmission electron microscope.) (continued on page 135)

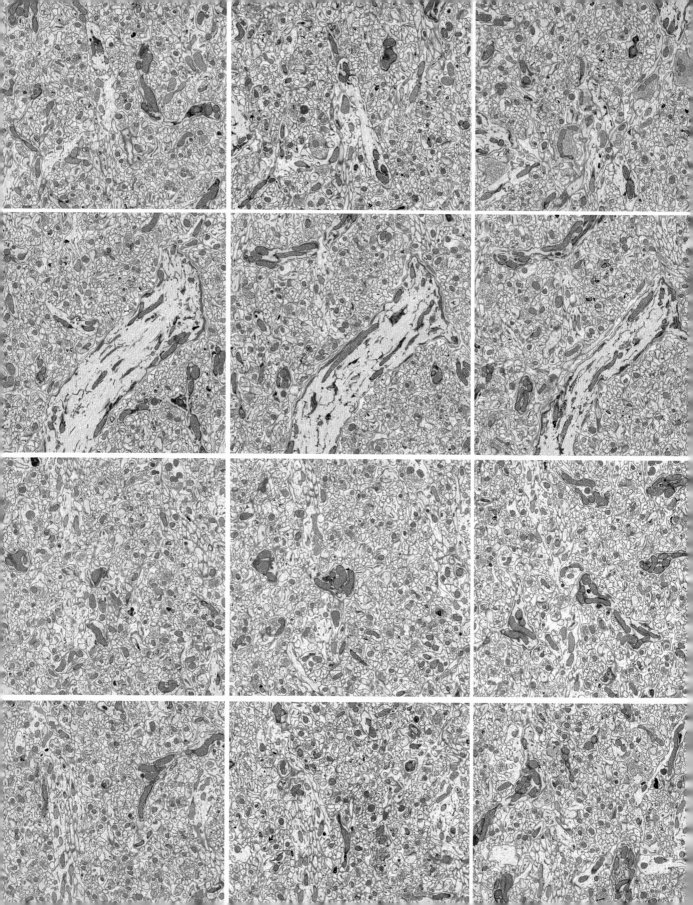

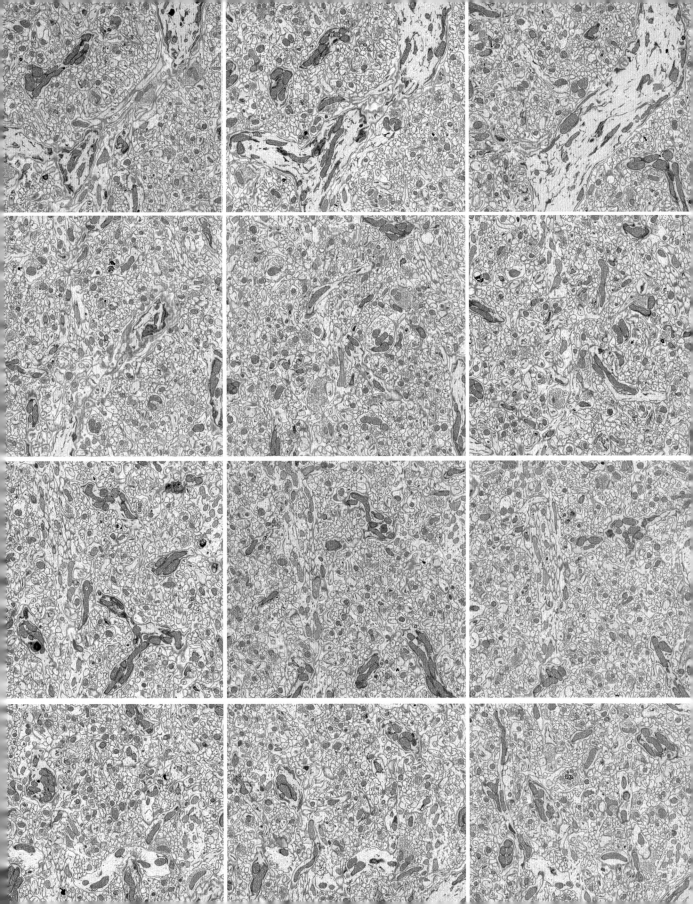

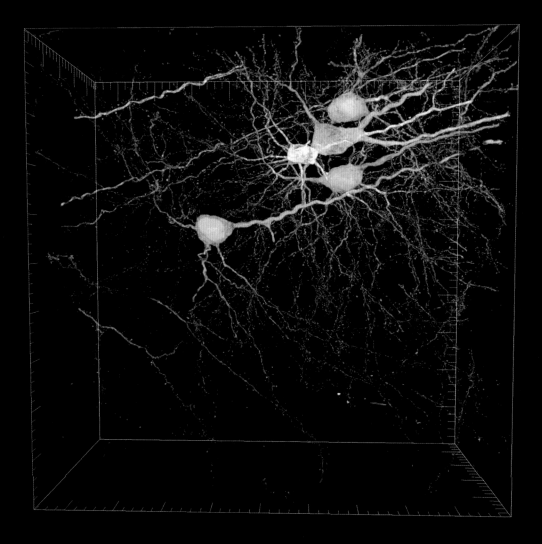

(continued from page 131) Once the surface of the sample is imaged, a sharp diamond blade automatically shaves off an exceedingly thin slice from the top. The microscope then takes another picture, the blade shaves off a bit more, and through this automated process the entire specimen is imaged all the way through, sixty nanometers at a time. The result is a series of naturally aligned, high-resolution images through the volume of the entire sample. The sections that appear on the previous spread are twenty-four of the hundreds of such pictures usually obtained in an imaging session. Each panel is a slice about six hundred nanometers from the next (arranged horizontally, from top left). The raw data alone offer little in the way of usable information and require careful analysis; in fact, one of the field's great challenges (discussed in more detail in Chapter 6) is to determine methods for making sense of the staggering volumes of data generated by serial block-face scanning electron microscopes. Many research teams are focused on developing computer algorithms that can read long series of two-dimensional images and use them to extract the three-dimensional structures of the neurons. The image opposite shows five individual neurons whose contours have been automatically traced by a computer from eight hundred consecutive imaged sections. The white cube enclosing them represents x, y, and z axes to convey a sense of depth.

POTASSIUM ION CHANNEL

Below: Model.
Opposite: X-ray diffraction pattern.
Stephen B. Long, Ernest B. Campbell, and
Roderick MacKinnon, 2005.

Some things are too small for even an electron microscope to see. Proteins are the molecules behind every event in the life of the cell, yet their structures remain beyond the electron microscope's reach. The family of proteins called *ion channels* give a neuron its electrical faculties and are plugged into its membrane, creating tunnels between the space inside the cell and the solution it bathes in on the outside. By allowing only specific, electrically charged ions (such as sodium or calcium) to travel across the tunnels they form, ion channels regulate the passage of electrical charges into and out of the neuron. Understanding the structure and function of our body's dozens of ion channels is a critical step to addressing severe pathologies of the nervous system directly linked to them, including cystic fibrosis, seizure, and certain forms of paralysis.

The image below shows the shape of a channel that allows potassium ions to pass through. The channel is seen "from above," with a view through its tunnel, where a small green sphere represents a potassium ion crossing from one side to the other.

These structures are exceptionally difficult to investigate; since researchers first revealed the structure of an ion channel in 1998, each addition to our knowledge of these diverse proteins has been hard-won. The difficulty resides in having to form crystals of the channel under study; like salt, proteins can arrange themselves into solid crystals that are laid out in a perfectly orderly pattern, given the right conditions. Once a channel is crystallized, researchers can shine a beam of X-rays through the crystal and, by measuring the way in which it is diffracted (the scattered pattern of dots in the image opposite), can deduce the original structure of the protein. This same principle enabled the discovery of the structure of DNA; to this day, it is the most powerful means not only of revealing the shape of biological molecules, but also of instantly and definitively closing the book on countless questions about their function. For in the world of proteins, as for brain circuitry, form determines function.

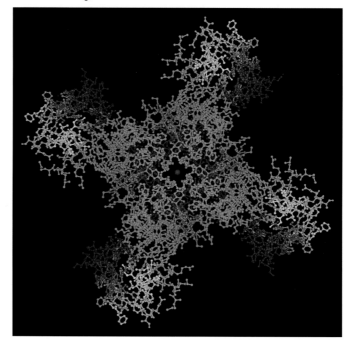

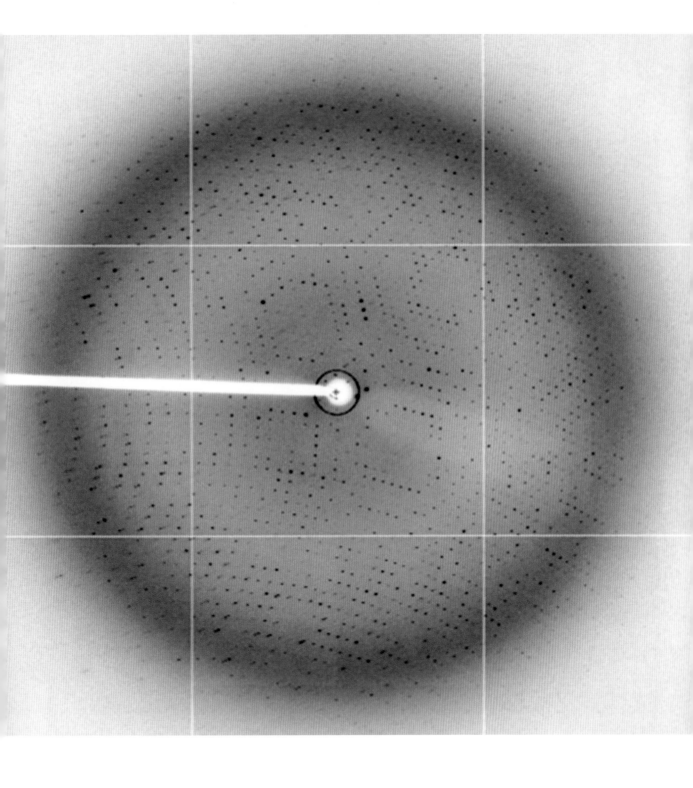

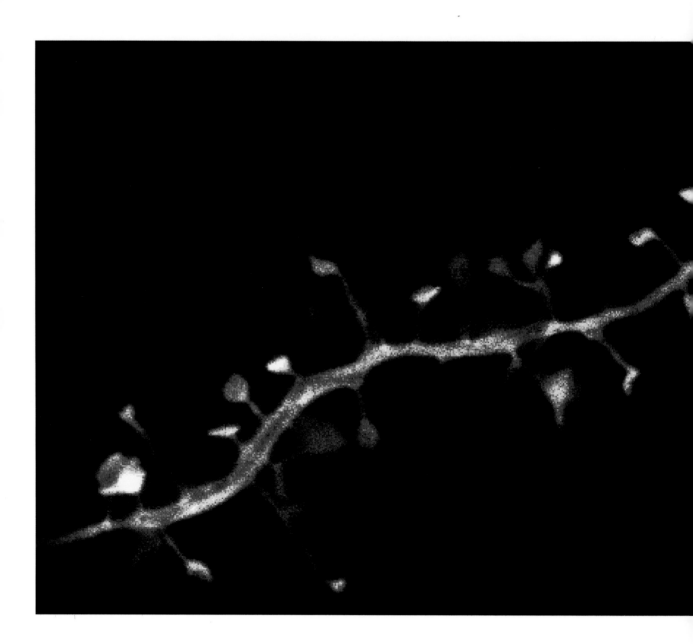

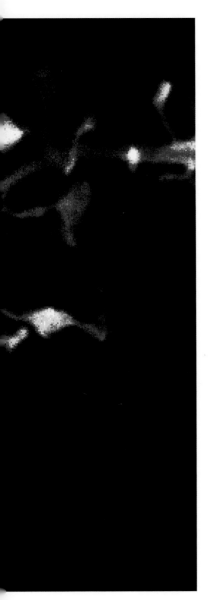

Nanoscopy: stimulated emission depletion microscope.
Katrin Willig, Nägerl Valentin, Nicolai Urban, Tobias Bonhoeffer, and Stefan Hell, 2008.

In microscopy, as in many other forms of scientific measurement, the higher the resolution, the more narrow the view. Light-based microscopy has been the standard tool for studying large areas quickly, but it is hobbled by the diffraction barrier, which sets a fundamental restriction on the size of objects it can resolve. Electron-based microscopy is not nearly as limited by this problem but is narrow in its scope, even though new methods like serial block-face scanning electron microscopy may expand it.

A very recent shift in thinking about light microscopy has resulted in the spectacular if improbable feat of using light microscopy to detect structures theoretically below the resolution limit of light. The diffraction barrier was "broken" not by cheating optics (physics is physics) but by shifting the focus of innovation. In a remarkable revolution in outlook, the microscopy avant-garde is now concentrating on exploiting the properties of fluorescent markers rather than on the optics employed to detect them. Several methods have appeared in concert over the past five years, resulting in a confusing alphabet soup of acronyms (STED, PALM, STORM, etc.).

Like a confocal microscope, a STED (stimulated emission depletion) microscope scans a laser beam across a sample; normally, this illuminates fluorescent molecules in an area roughly the diameter of the beam itself, but STED exploits the very workings of fluorescence and prevents all but those molecules in the very center of the laser beam from emitting light. In this way, the diffraction barrier of light is, paradoxically, "broken" using a light-based microscope. With it comes the potential to exploit the advantages of light (efficiency, scalability) while at the same time circumventing its blind spot.

As compared with the image of a dendrite's spines obtained using conventional microscopy (see page 119), the sheer clarity of STED images and its fellow nanoscopy prototypes have many in the field collectively holding their breath in the hope that these new tools, still in their infancy, will deliver on their tantalizing promise.

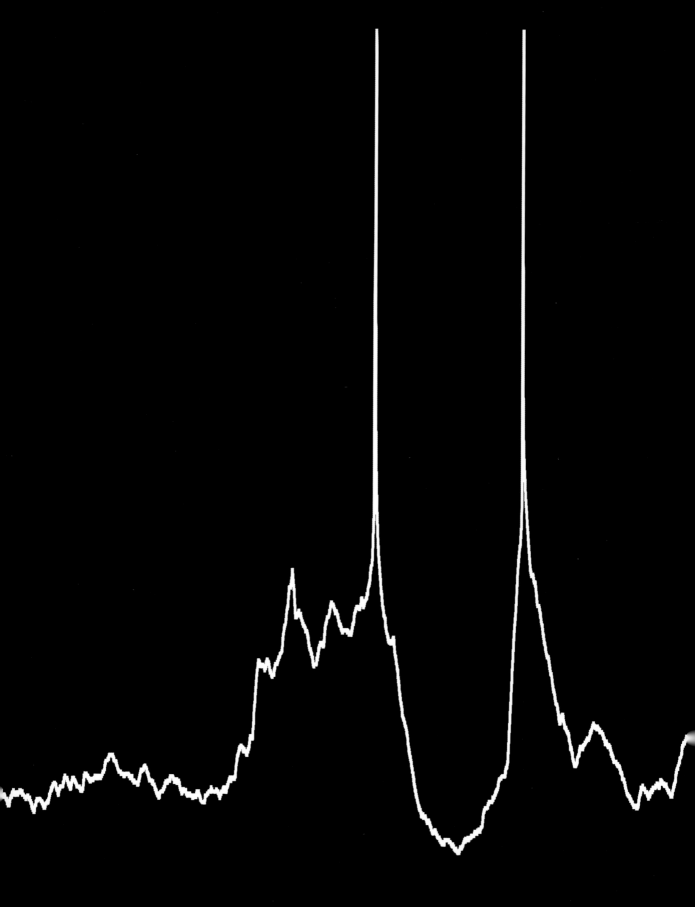

5.

ELECTRICITY IN THE BRAIN

ELECTRICITY IN THE BRAIN

Based on an interview with Michael Goldberg

Columbia University

Hippocrates knew in the fifth century BC that if you damage the left side of the brain, paralysis develops on the right side of the body. But it wasn't until the mid-nineteenth century that the examination of mind, brain and behavior took off in earnest, notably with the work of Paul Broca. A French physician, Broca had two patients who could understand language perfectly well but could not generate proper sentences. This was an astonishing situation in which the patients could understand without any trouble, had no problems with dexterity or with mouth movements, and yet could not speak or write.

The first patient was called Monsieur Tan because "tan" was the only word he could articulate. When Broca had the opportunity to examine Tan's brain after he died of natural causes, he found a lesion (severely damaged or dead tissue) in the left frontal convolution now known as Broca's area. The second patient had a lesion in the same place, and based on this evidence Broca was able to provide sound empirical grounds for the hypothesis that the cerebral cortex is divided into very specific areas that have very specific functions—in this case, language.

So it was neuropsychology—the study of how brain lesions relate to perceptual, cognitive and behavioral faculties—that enabled scientists to begin parsing how the brain functions. Sadly, what truly fueled the rise of modern neuropsychology were the Russo-Japanese War, and shortly afterwards, the First World War. Both conflicts produced a large number of patients who presented small, precise lesions in the brain caused by high-velocity bullets, which talented physicians like Tatsuji Inouye in Japan, and later on, neurologists in Great Britain, studied in order to piece together detailed functional maps of the brain. The logic behind these kinds of studies is straightforward: For example, a patient whose occipital lobe (an area in the back of the cerebral cortex) is injured in one of his brain's two hemispheres will be incapable of seeing one half of his normal visual field. From this observation, one may conclude that the occipital lobe is involved in visual processing.

✳ ✳ ✳

The notion that a physical part of the brain performs a particular function took on an entirely new significance in the 1950s when neurophysiologists began to insert very thin metal microelectrodes into the cerebral cortex and to record the electrical activity of neurons. Lesion-based neuropsychological studies, no matter how fine, rely on the irreversible damage of millions of neurons; thus, the conclusions they permit are fairly gross by necessity. At best, they can inform on whether a certain part of the brain is involved in a given behavior, sensory modality, or cognitive faculty. The next obvious question—one that neuropsychology alone cannot answer—is *how does it do that?*

As neurophysiologists invented new techniques to manufacture microelectrodes, they developed the ability to extract the lone voice of single neurons that would otherwise be lost in the ocean of neural murmurs that surround them. Armed with these new tools, investigating the brain in action became, for the first time, a study of how its individual neurons give rise to its properties. Thus, the anatomical resolving power that Golgi's method opened up in slices of dead brain tissue had finally found its functional expression, decades later, in the study of living brains.

But how exactly does one relate the activity of a single neuron to a perceptual, cognitive, or behavioral phenomenon? A neuron "fires" electrical action potentials (the basic unit of communication in the brain, described in more detail later in this chapter) at a certain rate even when it is not actively engaged. When its owner perceives a flash of light, memorizes an event, or moves an arm or a leg, if that neuron happens to be involved in the chain of events that underlie that faculty, the rate at which it fires will be altered. Thus, the fundamental theorem of neurophysiology in awake animals is this: If a neuron's firing rate tracks what the subject perceived or did, then that neuron is involved somehow in the analysis of the sensory environment or that behavior.

In practice, it can be difficult to find convincing applications of this theorem. In the late 1950s neurophysiologists David Hubel and Torsten Wiesel began recording from cells in the primary visual cortex in order to understand what makes them tick. Like other research groups that had tried and failed before them, they struggled to determine what visual stimuli elicit a response from these neurons:

> *In our very first experiments we used circular spots of light, or black spots. . . . But for the very first day or so we had no success in getting any clear [neural] responses . . . [Then] suddenly, just as we inserted one of our glass slides into the ophthalmoscope,* the cell seemed to come to life and began to fire impulses like a machine gun. It took a while to discover that the firing had nothing to do with the small opaque spot—the cell was responding to the fine moving shadow cast by the edge of the glass slide as we inserted it into the [ophthalmoscope's] slot.[1]*

This seminal experiment uncovered the fundamental property of neurons in the primary visual cortex: They detected lines in the visual scene but "completely ignored our black and white spots."[2] To this day, students of the visual system are still working out the implications of this momentous discovery.

✳ ✳ ✳

*An instrument most commonly found in an optometrist's office, which was tweaked in order to project images onto the retina in a precisely-controlled manner.

Until the 1960s, all the recordings were performed on anesthetized (sleeping) experimental subjects such as cats and monkeys. This is very convenient in studies of the visual system, for instance, as the experimenter can account for all of the details in the animal's field of vision: Since its eyes and head are motionless, it is possible to know with some precision what patterns of light fall onto its retina, and relate those patterns to neural activity. However, neurophysiological findings in an anesthetized animal may have little or no bearing on actual brain activity during consciousness. Moreover, a sleeping animal cannot learn, make decisions, or perform motor actions, so the range of questions available to the experimentalist working with anesthetized animals is quite limited.

Fortunately, there are no pain sensors in the brain, which allows one to record the activity of neurons using microelectrodes in awake, behaving animals without their experiencing discomfort. (Neurosurgeons take advantage of this feature when they operate on human patients, who are kept awake during the procedure in order to minimize the risk of damaging areas responsible for critical mental faculties.) It took several feats of engineering and behavioral training to develop the conditions that would allow recording in awake, behaving monkeys, but by the early 1970s the technique began to be adopted in labs around the country. And once the floodgates opened, some of the most enduring questions in neuroscience could at last begin to be addressed experimentally.

It was in the late 1960s in Robert Wurtz's young laboratory at the National Institutes of Health where everything came together. Using methods he and others had pioneered, Wurtz was the first to confirm that what Hubel and Wiesel saw in anesthetized animals also holds when they are awake. He then quickly moved on to investigate a fascinating, century-old hypothesis by the nineteenth-century physicist Hermann von Helmholtz: We—and presumably all visual animals—experience the world as being quite stable, yet as we move through our environment our bodies and heads are continuously in motion. Moreover, our eyes are constantly making fast movements (or *saccades*) as they collect information about different areas in our field of vision. Helmholtz postulated that the brain must somehow compensate for the motion of our bodies and eyes. One way to do this would be to send a copy of the commands that it dispatches to the muscles over to its own sensory areas. By comparing the sensory information flowing into those areas with the copy of the motor commands, the brain could cancel out the effects of movement and create the illusion of stability.

This was an attractive idea, but in order to test Helmholtz's hypothesis experimentally, Wurtz would need to uncover evidence for a compensating signal of this kind. To perform this experiment, he trained a monkey to hold its eyes perfectly still for a few seconds. Once the animal was proficient at this task, the effect of light patterns on neural activity could be studied in great detail, for a light stimulus could then be flashed on a screen in front of the animal in such a way that it would occupy a predictable area of its retina. Wurtz could then ask whether neurons in visual areas of the brain are able to distinguish between the motion of objects on the retina caused by the real motion of objects in the world, and the motion on the retina caused by the eye's own movements across a stable object. In an initial series of experiments in the primary visual cortex, he found that neurons are incapable of telling the difference between the images caused by world-generated motion and by self-generated motion across the retina. The results suggested that the problem must be solved in another part of the brain. Sure enough, the lab reported soon thereafter that a "Helmholtzian" signal could be found in the *superior colliculus* (a structure deep inside the brain and separate from the cerebral cortex), proving the century-old conjecture.

Wurtz's influential early studies demonstrated that we can ask deep cognitive questions with microelectrodes, and they preceded an avalanche of discoveries about the physiology of visual cognition, attention, and motivation, which still bears fruit today. The methods he helped pioneer granted to physiologists the freedoms anatomists had enjoyed for decades thanks to the groundbreaking work of Golgi and Cajal: The ability to cut through the clutter and investigate the properties of individual neurons.

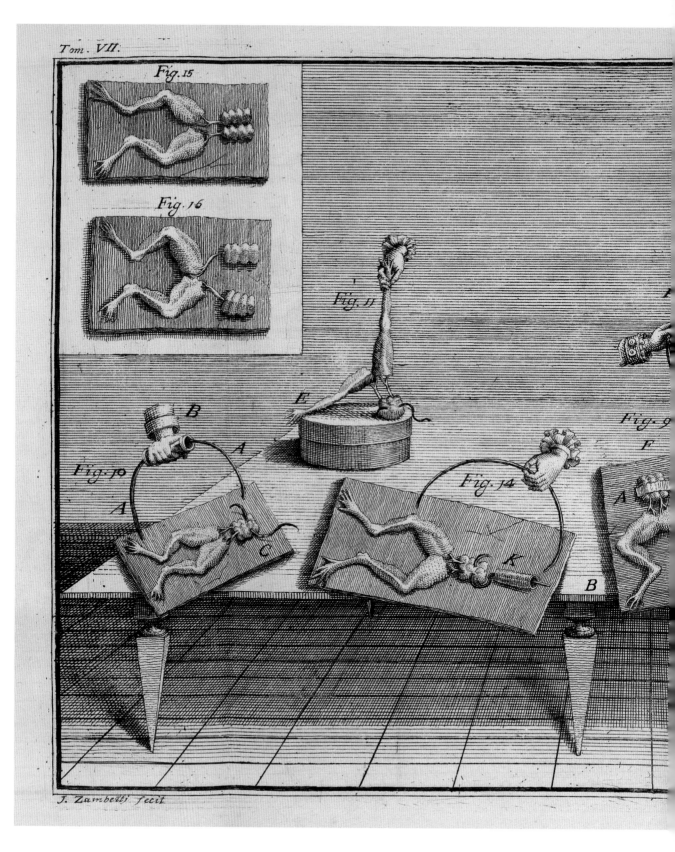

Early adventures in electrophysiology.
Luigi Galvani, 1791.

It is difficult to pin down the origins of electro-physiology, the study of the electrical activity in the body. In first-century Rome, patients suffering from gout were made to stand on live electric fish such as eels; people with headaches were instructed to apply the creatures to their foreheads. (Although useless as a remedy, perhaps the pain induced by this procedure masked the original discomfort.) The first formal understanding of electricity's role in the body is a relatively recent concept that dates from the eighteenth century.

Luigi Galvani, an Italian physician, most often receives credit for this paradigm, even if he and his wife and assistants weren't always the first to perform all the relevant studies. One in particular stands out, however. The story goes that the Galvani team's seminal experiment occurred when one of his assistants was fooling around with a machine that generated static electricity, while another assistant close by simultaneously touched a frog's sciatic nerve with a scalpel, thereby producing the momentous twitch that launched an entire field of study.

Severed, kicking frog legs were standard fare in the Galvani home laboratory. He describes the experiment labeled "Fig. 11" in this illustration, showing various procedures employed in the lab: "If a frog is so held in the fingers by one leg that the hook fastened in the spinal cord touches a silver plate and if the other leg falls down freely on the same plate, the muscles are immediately contracted at the instant that this leg makes contact. Thereupon the leg is raised, but soon, however, it becomes relaxed of its own accord and again falls down on the plate. As soon as contact is made, the leg is again lifted for the same reason and thus it continues alternately to be raised and lowered so that to the great astonishment and pleasure of the observer, the leg seems to function like an electric pendulum."[3]

149

More than two centuries after Galvani's antics, neuroscientists have expanded the study of electricity in the body to include our sensory experiences, our thoughts, and our emotions. Electricity is currency in the brain. A tightly choreographed ballet of electrical currents constantly—and fathomlessly—flickers throughout the vast expanses of the neural plains, engendering our every decision, every belief, every crush, and every aversion. Your experience of red in Warhol's *Campbell's Soup Cans*, your feeling of helplessness in the face of death, your body's reaction to a fall (along with its accompanying jolt of fear), and your most intimate secrets are all somehow carried by neurons that speak the language of electricity. Your very language faculty—learning English, deploying it in conversation, reading this book—is produced by their electrical activity. These neurons' currents were set into motion before you were born and will persist through sleep and consciousness until the final moments of your life. You are the summation of the electrical activity in your brain.

Neurophysiology, the modern field that sprouted out of Galvani's twitching frog legs, seeks to make sense of this electrical universe. The foundation on which the entire edifice rests is the diverse family of dozens of different proteins called ion channels (see page 136), which sit astride a neuron's membrane—the thin, oily "skin" that isolates the inside of the cell from the outside. Ion channels provide tunnels across the membrane, regulating the flow of electrically charged ions—such as sodium, potassium, and calcium—passing into and out of the cell. Each ion channel provides a discrete binary signal: either "all on" or "all off." When a channel opens up, ions enter or exit a neuron, causing it to become more positively or negatively charged and giving rise to the electrical fluctuations that carry information in the brain.

Ion channels open and close depending on specific physical events. They detect everything a neuron needs to know to function properly. Some channels found in the mouth, for example, have been shown to detect garlic (but only "raw, not baked," as one study found[4]); others translate the chemical events underlying synaptic transmission back into electrical signals; yet others are influenced by electrical events occurring inside the neuron itself. Each neuron contains a specific set of ion channels, which confers its particular identity: A neuron containing temperature-sensitive ion channels, for instance, is thereby a temperature-sensing neuron. Ion channels serve as the neuron's—and our—interface with the universe.

The image on the previous spread shows five sequential traces from a recording of a single human ion channel as it alternates between its open and closed states (respectively, down and up) in response to the presence of acetylcholine, a neurotransmitter. Each line represents four hundred milliseconds of time.

Obtaining a recording from a single ion channel is a delicate art. The revolutionary invention of the patch clamp, the method of choice for such recordings, opened up a universe by revealing these previously inaccessible signals from the minute electrical currents of single ion channels to the broad electrical fluctuations that occur in the entire neuron. To obtain a recording, a patch-clamp physiologist places a hollow glass cylindrical recording electrode directly opposite a very small patch of cell membrane—ideally containing only one or a small number of channels. A very tight seal is formed between electrode and membrane when the physiologist literally sucks on the small patch via a long piece of plastic tubing connected to the other side of the electrode. (Walk into any patch-clamp lab, and you will stumble on serious-looking Ph.D.s with tubing dangling from their lips like cigarettes.) The minuscule electrical signals captured by the electrode are amplified and displayed on a screen.

Above: Whole-cell recording in an awake rat.
Christine Constantinople and
Randy Bruno, 2009.
Overleaf left: Dual whole-cell configuration.
Greg Stuart and Bert Sakmann, 1994.

Here, researchers have employed an extension
of the patch-clamp method in which, after the tip
of their electrode is tightly sealed onto a neuron's
soma, they delicately but firmly rupture it with
a short, vigorous application of suction on the
tubing. This technique ("going whole-cell")
bestows upon the electrode access to the inside
of the neuron where it can spy on its inner
thoughts—the small, jittery electrical fluctuations that arise from the activity of countless
synapses, and which are invisible to an electrode
from outside the cell. If the single ion channel
recording on the previous spread were superimposed here at the same scale, the channel's
openings and closings wouldn't even be visible.

When these fluctuations reach a critical
threshold, the neuron emits an *action potential*—
a unit of information, the basic electrical signal
that allows neurons near and far to communicate
with one another. Two such events appear in the
trace above as sharp, vertical lines overwhelming
the small electrical fluctuations that occurred

below the activation threshold. The action
potential is the ultimate output of a neuron, a
summary of all the information it has collected
from its thousands of synapses; it travels down
the long axon and broadcasts, to all that care
to listen, that a neuron has spoken. When people
talk of brain activity, they usually mean these
action potentials, commonly referred to as
spikes, each one a carefully synchronized
sequence of ions gushing into and out of the
neuron in quick succession, giving rise to the
dramatic deviation from the small fluctuations
that preceded it. In this four-second recording
from the brain of an awake rat, each action
potential lasts approximately one millisecond.

With patience, effort, and skill, it is possible
to collect simultaneous whole-cell recordings
from several different parts of a neuron, as
illustrated in the image on the following page.
It shows two patch-clamp electrodes visualized
under a microscope where they take on a
triangular appearance. The one filled with the
blue dye records electrical fluctuations at
the soma, and the one filled with a green dye
records in the dendrite. Together, these can be
employed to reveal how local electrical events
that occur in separate parts of the neuron can
influence one another.

If action potentials are a neuron's means of transmitting information, then it follows that understanding the patterns of spike timing of individual neurons will provide insight into what they are doing. But in the brain, where countless tightly packed neurons are constantly firing away, it can be exceedingly difficult to eavesdrop on just one needle in the neural haystack. Different methods are available, depending on the kinds of questions a researcher seeks to answer, but practically all involve sliding a thin electrode into brain tissue, connecting it to a sensitive amplifier, and recording the signals that come out of it.

In these two images, an array of sixteen electrodes arranged along a pole that is scarcely thicker than a hair has been delicately inserted into a mouse's hippocampus, a center for learning and memory. Because each electrode is set a small distance away from its neighbors, this recording configuration can yield educated guesses about the source of the signals that it detects through a triangulation of sorts. The physiologist can then intuitively map the signals being recorded to specific components of the circuit's anatomy. The image on the following spread shows raw traces from these sixteen electrodes (indicated by horizontal lines) during spikes fired by six different neurons (indicated by multicolor vertical columns) identified by this statistical triangulation. In each column, the largest ripple—the unmistakable spike—betrays the vertical location of the cell that generated the activity, supporting the veracity of the neuronal identification. (In the green column, the recorded neuron is closest to the sixth electrode from the top; in the yellow, it is third from the bottom.)

The plot in the image opposite illustrates how one can statistically isolate the spikes of an individual neuron from those of its neighbors. Each cell's spikes will appear slightly differently on a recording: Some will be "larger," others "wider," and so on. (This variety in spike shapes appears clearly on the next spread.) Any two properties can be plotted in a coordinate system like the one above, where each dot represents a single spike recorded from an electrode during an experiment; here, four clusters of spikes appear clearly, suggesting that in this recording the neurophysiologist was listening in on four different neurons. The data in these two images was obtained from experiments investigating how newly born neurons influence neural activity in the hippocampus as they are integrated into this circuit, a phenomenon that has potential implications for the action of antidepressants.

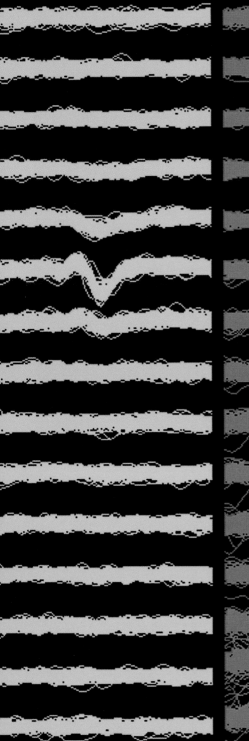
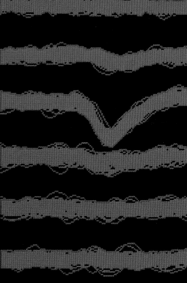
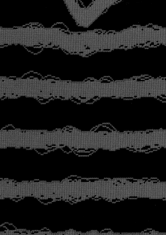
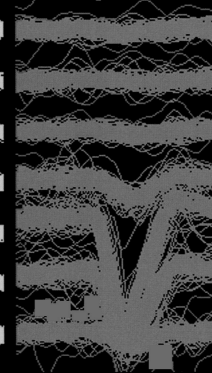

Action potentials in a live monkey brain.
Alexandre Saez and C. Daniel Salzman, 2009.

Spike recordings from single neurons have
yielded immense insight into brain function
and are among the most powerful means at
our disposal to decode its language. This example
shows a recording obtained from a neuron in
a monkey's amygdala—an area that tracks
positive and negative experiences. Each of
the thirty-four horizontal lines represents one
four-second-long trial, and each small vertical
notch indicates a spike. Two seconds into each
trial (in the middle of this figure) the monkey
was offered a reward, resulting in a sharp
increase in the number of spikes immediately
thereafter. Thus, this neuron can be said to be
signaling positive experiences. Any system
that is built to learn must be able to detect
pleasurable and aversive events; once these
neurally encoded signals are broadcast to the
entire brain, it can set about figuring out how
to obtain more of the things it likes and avoid
those it doesn't.

Current source density analysis.
Clay Lacefield, 2007.

In addition to obtaining recordings from individual neurons, neurophysiologists can study their coordinated behaviors across large expanses of the brain; what they lose in precision (the moment-by-moment firing patterns of individual neurons) they gain in breadth, providing insight into global patterns of activity. The plot at right shows a current source density analysis of signals obtained using the same electrode configuration discussed on page 153. This plot reveals the aggregate activity of an untold number of neurons in the brain of a sleeping mouse, as they fire synchronously, spreading currents throughout the entire space (about one millimeter) detected by the sixteen electrodes. The raw recordings from the electrodes are represented by the sixteen thin horizontal lines traveling across the image. An algorithm to analyze them reveals the positively charged "sources" of these electrical currents (the darker areas in the background) and the negatively charged "sinks" (the lighter areas corresponding to synaptic input) to which they travel, representing the underlying neuronal activity that gave rise to them. (Gray represents areas that are neither positively nor negatively charged.) Massive, all-encompassing events like the one that arises in the left third of this recording occur every few seconds in the sleeping brain. Researchers believe that they trigger the replay of events the subject experiences while awake.

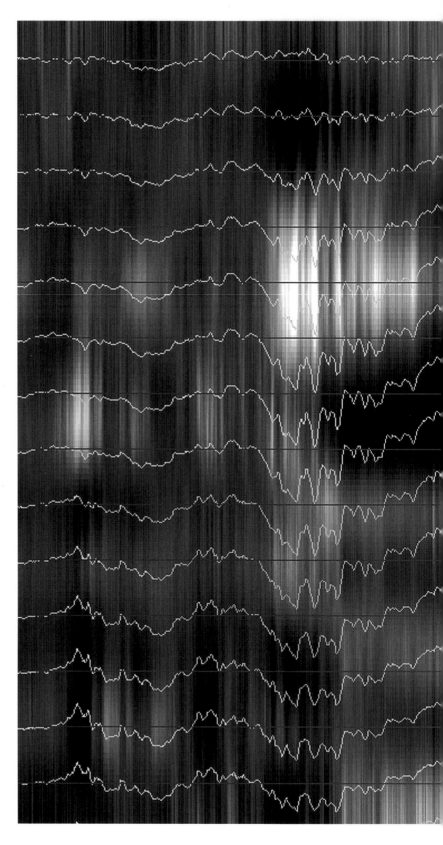

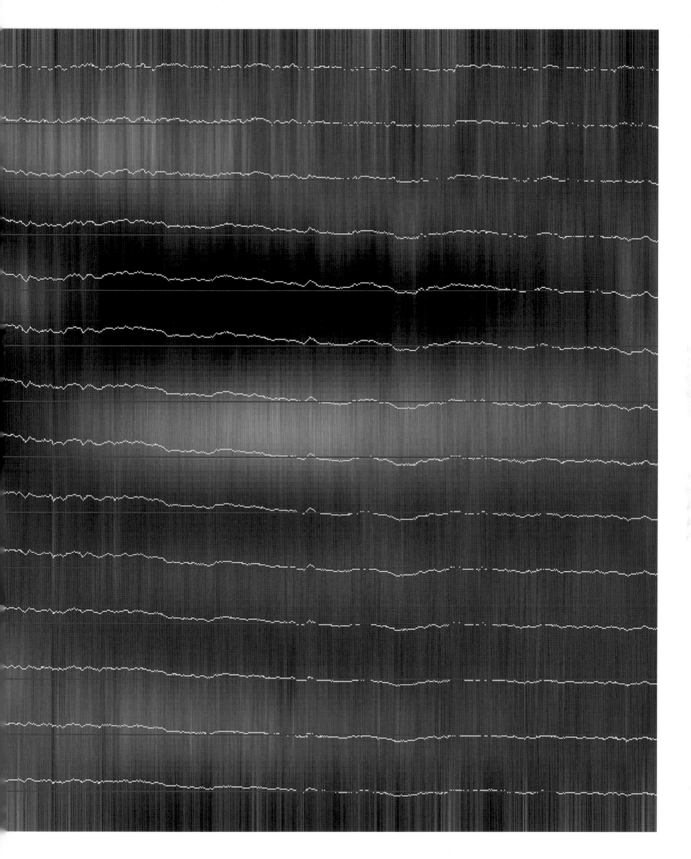

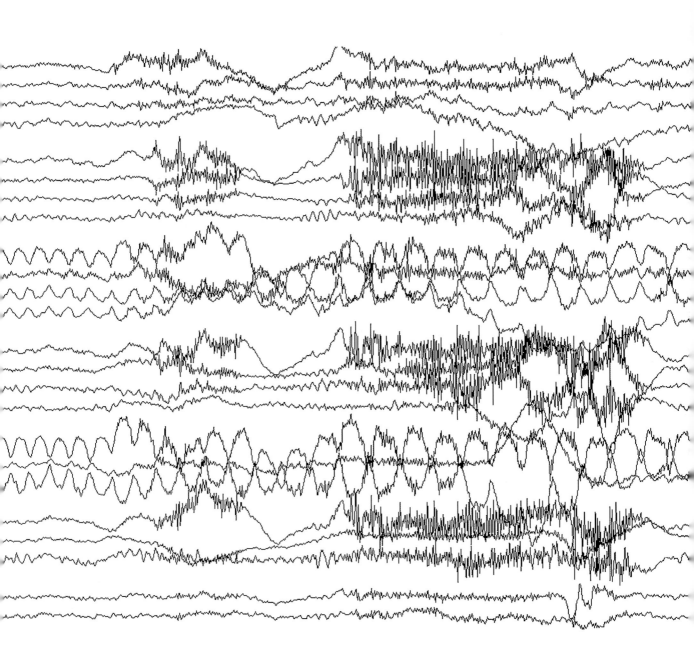

Electroencephalography: epileptic seizure.
Edgar Toro, 2009.

It is impossible to detect the spiking of neurons inside a human brain without cutting a hole in the head and inserting an electrode, but less invasive recording techniques have been developed. Large volleys of coordinated electrical activity can be detected using electroencephalography (EEG), which involves placing an array of noninvasive electrodes around the head. Since our bodies are composed largely of water, electrical activity generated in the brain is conducted through the skull and can be detected using these metal sensors. The signal is countless orders removed from the openings and closings of the single ion channels that actually generate the currents in the first place, but gross patterns can nonetheless be revealed through careful data analysis.

EEG is an ideal diagnostic tool for dramatic events such as seizures, which appear quite unambiguously. Like earthquakes, certain kinds of seizures start in one central location and propagate across the surrounding area. EEGs are routinely used in clinical settings to assist neurosurgeons in determining which region of brain tissue lies at the epicenter. Patients are fitted with EEG electrodes and placed under observation in a hospital for a few days. This fourteen-second recording shows the onset of a partial seizure in a patient fitted with twenty-seven electrodes that monitor the activity in the brain area directly beneath them. Initially, the traces are more or less flat, except for small fluctuations that register only on a handful of electrodes (those closest to the source of the seizure). They then spread across and induce violent waves that can be recorded across the entire hemisphere. By examining which electrodes exhibited abnormal activity before the others, clinicians can determine what area in the brain is initiating the seizures. They can follow up by implanting a few electrodes—this time, surgically introducing them through the skull—to narrow down the exact location of the offending tissue.

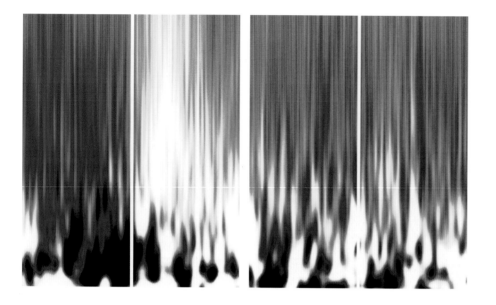

Time-frequency diagram.
René Quilodran, Marie Rothé, and
Emmanuel Procyk, 2007.

Time-frequency diagrams are often used to represent signals from electrode recordings after these have been analyzed by algorithms like the wavelet transform and the Fourier transform. These mathematical tools can decompose an electrophysiological recording into its component frequencies, revealing the rhythms that occur in the brain while it is engaged in various tasks. As the volume and complexity of available physiological data expands, researchers are increasingly turning to sophisticated algorithms to aid them in their data analysis and extract meaning from their recordings. It is especially critical to employ this kind of mathematical treatment when studying signals that are far removed from the neural events that give rise to them, as with EEGs. This time-frequency diagram was obtained from electrode recordings in the anterior cingulate cortex of a monkey, which were decomposed into their component frequencies with a wavelet transform. During these recordings, the monkey was playing a computer game in which it would try to guess where a reward was hidden by tapping different cues on the screen with its finger. Once it had figured out the sweet spot, the animal quickly learned to select that cue every time it encountered the same situation, thanks in part to the neural mechanisms evidenced by this data.

The flamelike surges betray the presence of specific brain rhythms, and in this experiment suggest that the brain area under study encodes feedback that an animal can use to explore or exploit a particular situation. Time (four seconds) runs from left to right; the relative power of a given brain rhythm (plotted on the vertical axis from 20 to 150 Hertz) is encoded on a spectrum that runs from dark red (low -110 microVolt^2) to bright yellow (high +110 microVolt^2).

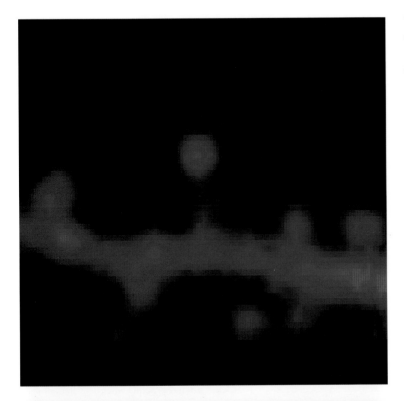

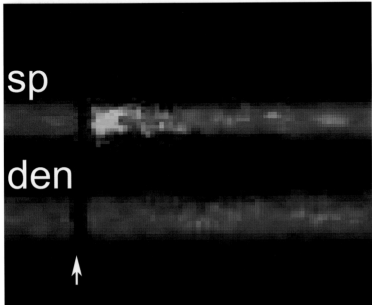

Two-photon calcium imaging of synaptic transmission.
Brenda L. Bloodgood, Andrew J. Giessel, and Bernardo L. Sabatini, 2009.

It is tempting to compare the advances made over the past two decades to those that occurred during the late-nineteenth-century golden age of anatomy, which followed the simultaneous introduction of novel microscopy and staining methods. The advent of two-photon microscopy in the 1990s (see page 121) and the recent development of molecular tools to finely record and influence neurons' electrical activity promise to revolutionize the practice of neurophysiology.

In the example shown here, researchers imaged calcium ions gushing into a neuron's spine through ion channels in its membrane when it was activated by the neurotransmitter glutamate. This process, which mimics synaptic transmission, is inaccessible to even the finest electrode-based neurophysiology methods, and here was studied using a two-photon microscope in combination with a fluorescent calcium sensor that lights up in the presence of that ion.

The top panel shows a dendrite with a few of its spines on either side. The panel beneath shows the level of calcium over time in the top spine ("sp") and the dendrite ("den") that it is attached to upon delivery of neurotransmitter (arrow). When the synapse is activated, calcium immediately flows into the spine, as indicated by the rapid increase in green signal, followed by a return to baseline—all within approximately forty milliseconds (under these conditions). An extraordinary view of synaptic transmission, this study reveals that despite the marked increase in local calcium ions in the spine, their flow does not appear to spread to the dendrite.

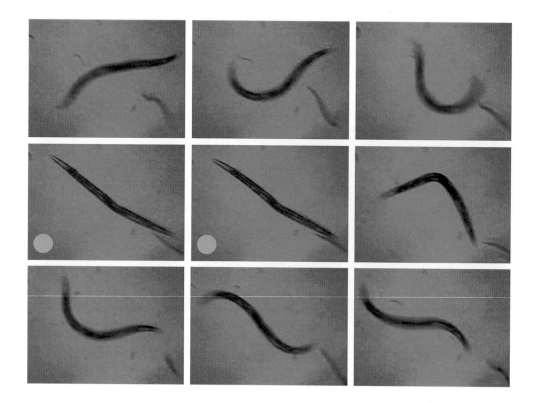

Channelrhodopsin in Caenorhabditis elegans.
Martin Brauner and Alexander
Gottschalk, 2006.

While electrodes can provide correlative information about how a neuron responds in a particular situation, it is very difficult to prove that it is causally linked to a particular perception or behavior. A fully developed science must be able not only to establish correlative information, but also to provide a causal mechanism for the system that it studies.

Thus, the development of light-sensitive molecules into agents that can control neural activity at will is potentially one of the most important in this past decade. The first and most famous of such molecules to date, *channelrhodopsin* and *halorhodopsin*, promise to overcome the long-standing causality challenge in neurophysiology.

The genes for channelrhodopsin and halorhodopsin—originally discovered in algae and bacteria, respectively—can be introduced into any neuron, where their resulting proteins are inserted into its membrane. There they can control the flow of charged ions into and out of the cell, thus influencing its electrical activity. Roughly speaking, channelrhodopsin, an ion channel, can switch a neuron "on" when activated by blue light, causing the cell to fire a spike; halorhodopsin, conversely, can be used to silence a neuron's output when activated with yellow light.

This fine control over the activity of neurons allows researchers to probe the effects they cause in a given network. Although this research program is still in its infancy, many believe it will prove to be as transformative as the discovery and application of the Green Fluorescent Protein (see Chapter 3).

This image shows one of the earliest experiments to demonstrate the power of channelrhodopsin to resolve causality, here in the worm *Caenorhabditis elegans*. The gene for this channel was delivered to one class of neurons that prevents muscle contraction when activated. When researchers switched on a blue light (seen in the two panels marked with a dot), the constantly writhing worm's motions are momentarily frozen. Today a freezing worm, tomorrow a cure for Parkinson's or epilepsy.

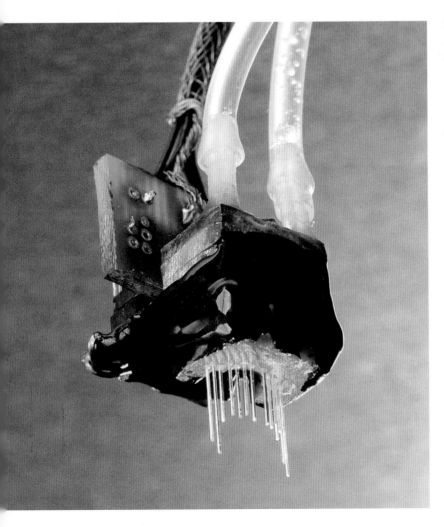

Optical fiber array.
Jacob Bernstein, Alexander Guerra,
and Ed Boyden, 2010.

Researchers reported using channelrhodopsin to partially restore vision to blind mice in a 2008 experiment. In 2009 another group of researchers was able to mitigate symptoms of Parkinson's disease in mice by activating channelrhodopsin in motor-cortex neurons. With the imaginations of neuroscientists the world over running wild following the discovery of these molecules, the field is faced with serious engineering challenges—most critically, how to target light beams onto the right neurons. Parallel to strategies aimed at improving on the natural forms of these molecules is the development of increasingly sophisticated means to deliver light to just the right structures in brain. One such device, the miniature array of independently controllable LED-coupled optical fibers shown here, is light enough to rest on top of a mouse or rat's head. This enables researchers to deliver light through its thin optical fibers to specific subsets of neurons in the animal's hippocampus. The goal is to causally investigate this brain area's mechanism with ever-increasing precision by selectively switching different parts of it on or off. By the time this book reaches store shelves, the inventor of this device will undoubtedly have improved on it several times over.

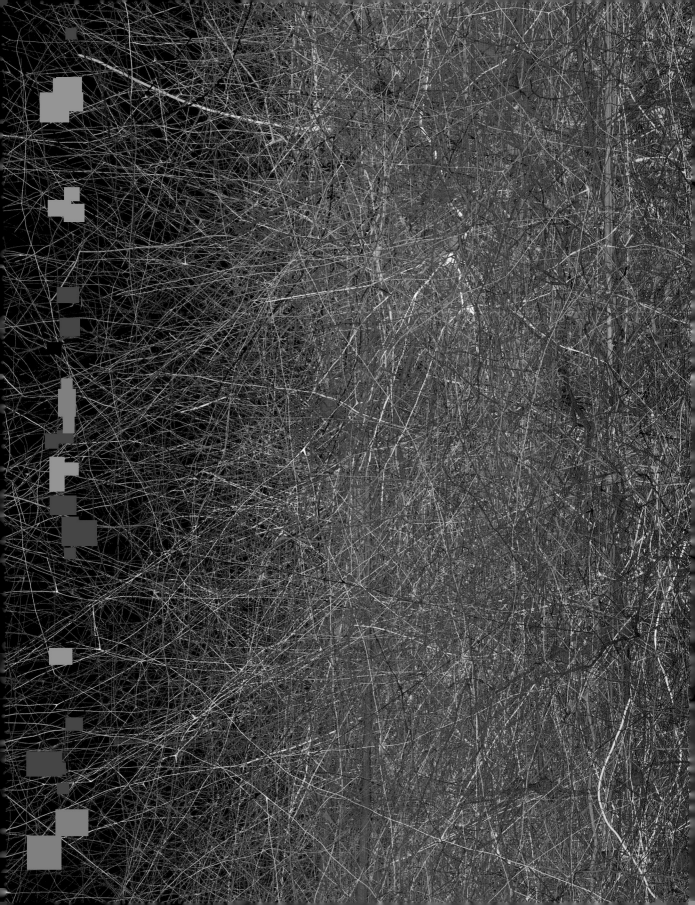

6.

THE BRAIN AS CIRCUIT

THE BRAIN AS CIRCUIT

by Terrence Sejnowski

Salk Institute for Biological Studies

Although Santiago Ramón y Cajal is best known for his magnificent drawings of neurons from the brains of many species, it is his insight into brain function that may be his greatest scientific achievement. A close look at the images on pages 67–71 reveals a sequence of small arrows that were obviously not present in the biological samples he examined. He added these arrows to his diagrams to indicate the direction in which he had deduced neural signals progressed within each neuron—from dendrite to soma to axon—and between neurons along their long axonal projections. In the process, Cajal established the overarching conceptual framework under which we approach the brain—the circuit. His little arrows were the germ of a theory of how information flows in the nervous system, a concept we are still trying to grapple with one hundred years after his seminal work.

While Cajal was developing his ideas about neurons and neural circuits, others were investigating brain function by examining the cognitive deficits associated with injuries to various parts of the brain. For example, focal injury to a region in the brain known as the motor cortex (an area involved in voluntary movement) produces weakness or even paralysis in muscles whose control depends on brain tissue at location of the injury. Patients with injuries to one area of this strip of cerebral cortex might have paralysis in one arm, while those with injuries to another area might have paralysis in a leg. These kinds of observations helped convince scientists of the hypothesis that mental faculties are localized—that is, that certain areas of the brain are responsible for certain functions. But knowing what goes wrong when some part of the brain is damaged is not tantamount to knowing that circuit's function; pinning a broad label like "vision" or "motor" on a brain region is a poor substitute for understanding how it works.

✳ ✳ ✳

When the late Francis Crick shifted his research program from molecular genetics to neuroscience in the 1970s he channeled his efforts on vision, studying how the connections between neurons in the visual system are organized. Neuroanatomy served as a conceptual framework for his thinking about higher-level processes such as visual awareness and consciousness. The dictum that had served him well in deciphering the function of DNA from its double-helical structure ("from structure follows function"), also offered powerful clues about the function of neural circuits. If a circuit receives input from a visual area, for instance, it is very likely that it in turn exploits or further processes visual information. Crick was particularly intrigued by recurrent loops between brain areas, as he thought that they might have something to do with our ability to pay attention to objects in visual scenes. He conjectured that the feedback connections between neurons in the visual cortex and neurons in parts of the thalamus—a brain structure that sends inputs to the visual cortex—might regulate the incoming information and underlie conscious experience.

But in order to fully account for the anatomical underpinnings of brain connectivity, we need more than just the broad picture of how one major area communicates with another. The Golgi stain reveals the shapes and sizes of neurons in every part of the brain and, to a limited extent, can help researchers deduce the patterns of connectivity between them, but in order to fully understand brain function we will require a complete wiring diagram of both the connections between all the neurons in a single brain area—the local circuit—and the long-range connections between different brain areas. Just as genomics made it possible to compile a complete list of genes and compare them between species, the new field of connectomics seeks to compile complete circuit diagrams of various nervous systems, including the human brain. The wiring diagram of *C. elegans* with its 302 neurons (see page 186) was recently constructed with painstaking human labor, but it will soon become possible to piece together diagrams of much larger systems using powerful computational techniques that automate the process. How will these new maps of connectivity change the way we think about brain function?

A circuit diagram by itself does not reveal a circuit's function, as a diagram is inherently static and brain function is determined by dynamic processes. Today's equivalent of Cajal's little arrows are computer simulations of brain models that take into account the anatomical details and the biophysical realities of neurons and the synapses that connect them. The goal is to follow information as it enters the circuit, such as patterns of light hitting the retina in the eye, and to investigate the resulting signals as they propagate down a chain of neurons, circulating through recurrent loops inside the brain. This is a daunting task since many details about neurons remain unknown, but also because even the fastest supercomputers cannot keep track of all the billions of brain cells that make up a human brain. The system is simply too complex for our computers to fully simulate with today's technology.

When I first began using computers to model neurons and neural circuits in 1980, I used computers that could perform around a million operations per second (at the time this seemed enormously powerful, but today I would be better off using a smartphone). With this technology I was able to simulate a few hundred simplified neurons, each connected to many others. Even using just a small number of these virtual "neurons" I was able to demonstrate how artificial "neural networks" could be configured to perform some amazingly complex tasks, such as pronouncing English words. Rather than hand-wire the network using rules for English pronunciation, which have many exceptions, I created a network trained on examples. Every time the network made a pronunciation mistake, the strengths of "synapses" between "neurons" were changed by a small amount so that the next time the same word or similar words occurred, the pronunciation would improve. This was a slow process. At first the network babbled like a baby as it learned to distinguish consonants from vowels. But eventually it began to sound out simple words and finally even complex words with many syllables. This was astonishing to me and others, who up until then had believed that English pronunciation was beyond the reach of a simple neural-network model. The experiment taught us that problems that can be difficult to program a computer to solve can be quite simple for a network to learn from experience.

Today's computers are thousands of times faster than the ones I used in the 1980s, and we can now simulate neurons in much greater detail, including the fine branching of dendrites and the many synapses on dendritic branches. Theoretical neuroscientists have employed computers to simulate the geometry of the neurons that Cajal first observed, revealing that a single neuron is far more computationally powerful than previously thought; even though the brain derives many of its remarkable abilities from the connections between neurons, some of its prowess is due to the intrinsic dynamical properties of the neurons themselves.

✻ ✻ ✻

If the study of neural circuits weren't sufficiently complicated, it is now known that circuits are dynamic on many timescales. Every protein in every cell is replaced over a matter of hours or days. Synapses between neurons are plastic and can change their sizes and strengths in response to changes in the patterns of activity within circuits, and new synapses are formed on a daily basis. Massive synapse remodeling occurs during brain development and continues at a reduced rate in adults as a consequence of learning through experience and, sometimes, recovery after brain damage. The loss of neurons due to age or disease leads to a deterioration in memory and agility, another structural shift within the brain that affects neural circuits. In order to fully address the challenge posed by this constant flux, researchers must map many circuits at different stages of development and in many different environments.

Yet, even if we could reproduce all the anatomical details and signals in a brain, this wealth of knowledge would not in itself explain how a brain functions, or goes awry. What we need is a twenty-first-century Cajal who can understand the function of these circuit diagrams by simulating the signals as they are processed by the circuits themselves. For example, while a person is looking at a particular object, the neurons that represent that object in the brain change their firing patterns, sending out signals to other neurons at faster or slower rates. This is a subtle change that does not involve a shift in the wiring diagram of the brain; we will need to quantify these firing patterns using theoretical tools like information theory from engineering and dynamical systems theory from physics. But even this may not be enough to fully understand brain function. There are many complex systems in the world that have defied our best efforts to understand them both through mathematical analysis and simulation. The weather, to pick a familiar example, remains famously unpredictable despite decades of research, sophisticated mathematical models, and ever more powerful computers.

✳ ✳ ✳

It may be that in order to fully understand brain function we must first understand how the brain develops from an embryo, and how the molecular mechanisms inside cells interact with the information flowing through brain circuits. In particular, we need to resolve how structures within the brain at many different spatial scales interact with each other over a wide range of timescales, a challenge termed "the levels problem." For example, the relatively slow molecular machinery inside cortical synapses, which regulates the size and strength of a synapse, is itself controlled by the relative timing, on a very short timescale, of the action potentials in presynaptic neurons (those that are sending messages) and postsynaptic neurons (those that are receiving them). The patterns of action potentials in a population of neurons can influence the biochemical pathways inside synapses that control the way neurons communicate with each other, which, in turn, influence the subsequent pattern of action potentials. It is with this elegant interplay between scales that the brain is able to solve the countless variety of problems that it is designed to tackle, as well as challenges that nature could not anticipate. And it has proven a source of constant frustration to students of the brain who are attempting to untangle its function.

We are witnessing a period of unprecedented innovation in the techniques for studying the brain. The hope is that just as Golgi's method enabled powerful insight into the structure of the brain, today's dazzling new tools will open up a universe hitherto inaccessible to us. And indeed, they have begun to bear fruit: we can now "watch" the electrical activity of many neurons simultaneously using optical recordings; the advances (described at the end of Chapter 5, pages 164–165) even allow us to manipulate the activities of individual cells and cell types using beams of light. When new techniques are introduced, it is more or less impossible to predict what will come out of them, or whether they will even alter a field's landscape. Golgi's technique lay fallow for more than a decade before Cajal perfected it and applied it to establish the basic facts about neurons. We can only hope that another Cajal will soon emerge to fully exploit this embarrassment of riches and uncover the secrets of our marvelous neural circuits.

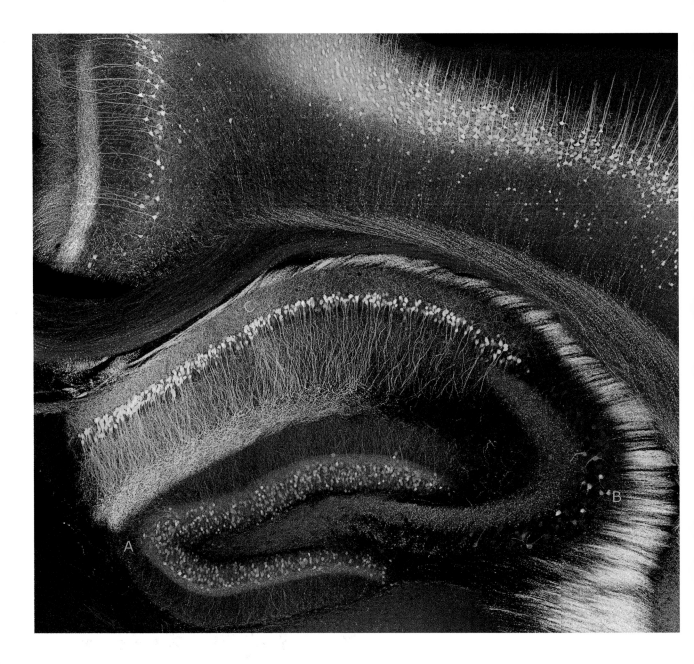

HIPPOCAMPUS

Opposite: Broad overview.
Tamily Weissman, Jeff Lichtman, and
Joshua Sanes, 2005.
Overleaf: Subnetwork.
Gyorgy Buzsáki and Attila Sik, 1995.

The late neurological patient known as "H. M."
provided invaluable insight into human cognition
when his debilitating epilepsy was treated half
a century ago by the surgical removal of large
portions of his hippocampus. While cured of
his seizures, he also lost his ability to form
long-term memories. Astonishingly, he could
acquire and deploy new motor skills yet could
not recall having learned them. His plight yielded
compelling evidence for the view that different
forms of memory are handled by different areas
of the brain—that specific circuits, such as the
hippocampus, serve specific functions like
conscious memory.

This image and the one on the following
spread show cross-sections of a mouse's
hippocampus, revealing its intriguing architecture. It is shown in the bottom half of the image
opposite, nestled directly beneath the neocortex.

The neurons in the hippocampus (whose
somata appear here as small circles) can be
grossly divided into three regions, defining the
sequence in which information flows through
the hippocampus. Most of its input arrives from
the *entorhinal cortex* (to the left, not pictured),
delivering information from many different
sensory systems, into the *dentate gyrus* (A),
through the *CA3 region* (B), and up to the *CA1
region* (C). The circuit then closes the loop as
CA1 sends its output back to the entorhinal
cortex. Of course, in reality the network is far
more complex: For instance, the entorhinal
cortex actually projects to all three areas
independently, and the CA3 region exhibits
a rich interconnected subcircuit within it.

Indeed, these broad anatomical subdivisions belie the fact that each area in the
hippocampus is itself an intricate network of
thousands of individual neurons. The image
on the following spread shows a single neuron's
soma and dendrites (at center, in orange) and
the dense branches of its axon (in yellow)
spreading throughout the entire dentate gyrus.
The overall structure of the hippocampus is
outlined in the background in blue.

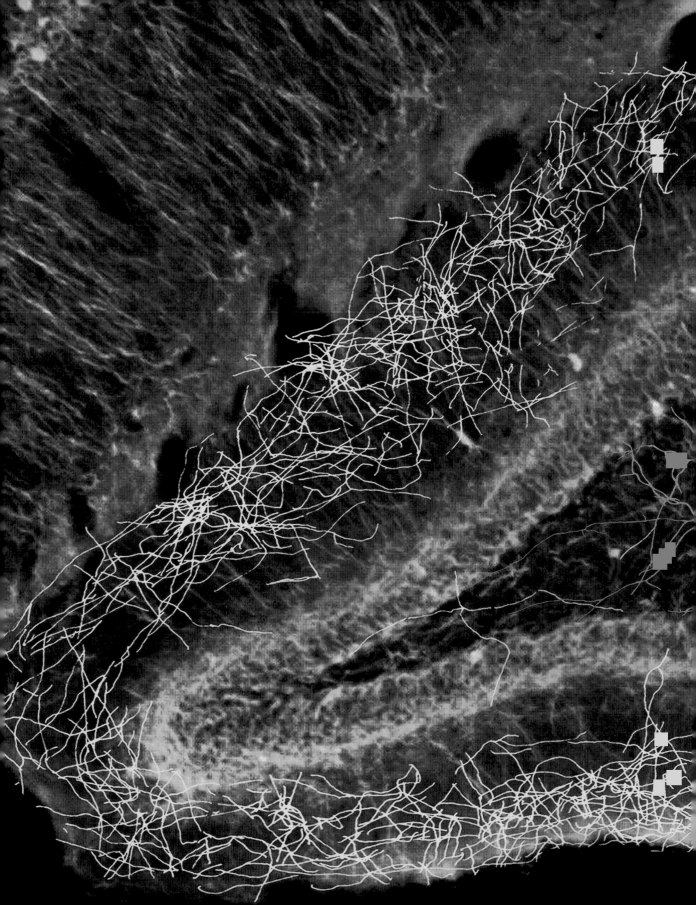

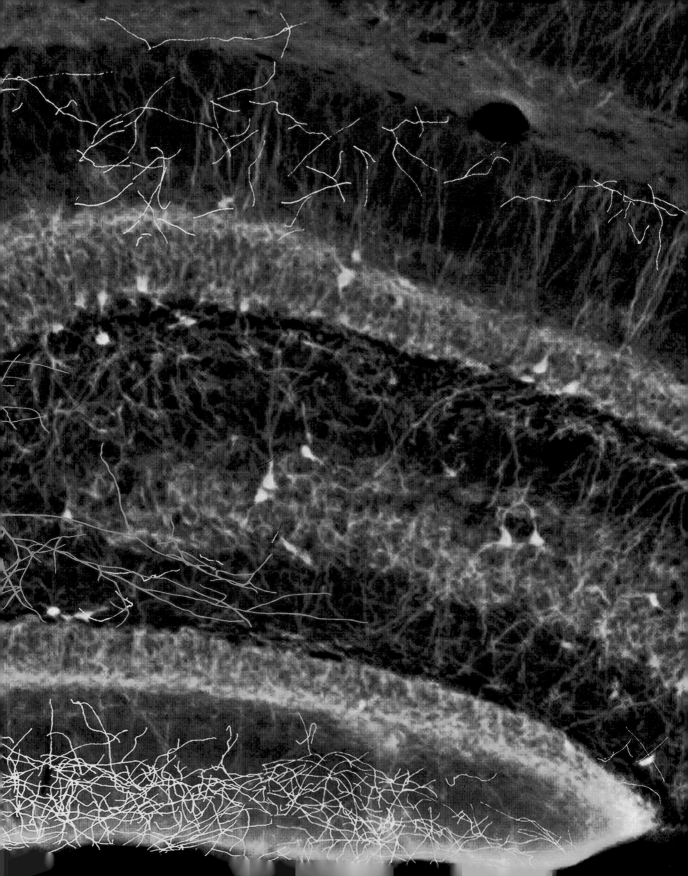

Neocortex.
Tamily Weissman, Jeff Lichtman,
and Joshua Sanes, 2007.

This image zooms in on an area of the neocortex
(seen from afar on page 172) to reveal a
horizontally layered organization: Behind the
colorful somata in the foreground, a pattern of
light and dark in the background suggests
anatomical distinctions. The darkest band in the
upper portion of the image is the first to receive
the bulk of sensory input from the outside world,
then (unmarked) neurons in the region pass
the information along to the neurons above and
below for further processing and broadcasting
to far-flung areas in the brain.

This type of horizontal layering is a
general organizational principle of the cortex;
it is found—with variations—in areas that
underlie seemingly unrelated tasks such as
processing visual information, planning motion,
and making rational decisions. The details of
how these complicated cortical circuits function
are under intense investigation, in part because
our oversize cortex sets us apart from other
mammals. It is so large that it was forced to
fold upon itself in order to fit into the skull as
our brains expanded over the course of evolution.
Thus, understanding the cortex may be the key
to understanding how the brain makes us who
we are.

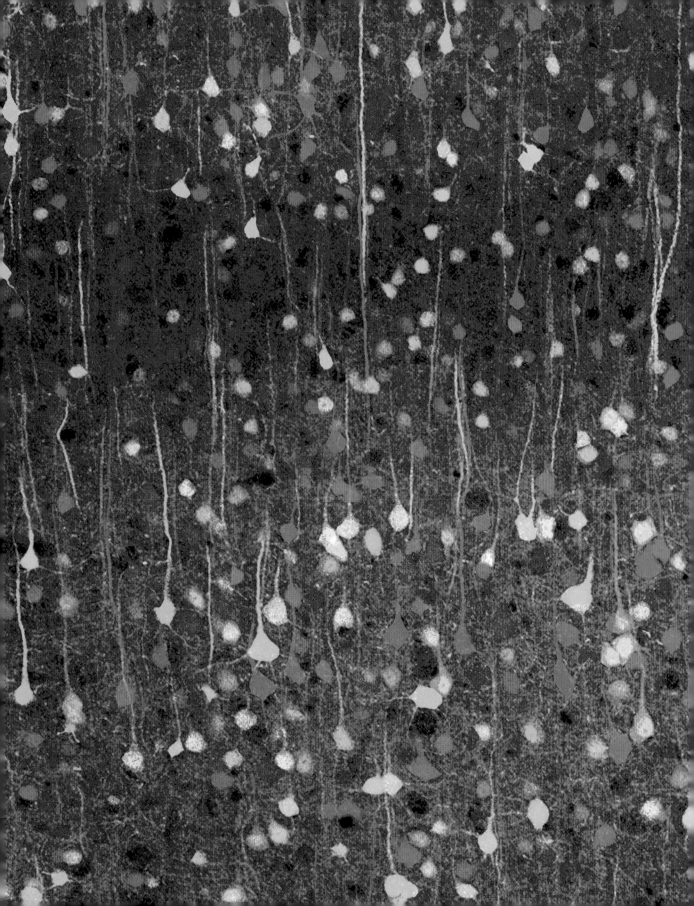

Olfactory sensory neurons.
Peter Mombaerts, Fan Wang, Catherine
Dulac, Steve Chao, Adriana Nemes,
Monica Mendelsohn, James Edmondson,
and Richard Axel, 1996.

With this image, we travel out of the brain and
into the inner surface of the nose. Here, *olfactory
sensory neurons* give rise to our sense of smell
by detecting scent molecules floating through
the air—like those wafting out of a bakery
carrying the smell of fresh bread. These molecules
bind to receptors embedded in the sensory
neurons' membrane, translating the molecular
reality into signals understandable to the brain.
Although there exist hundreds of different genes
for these receptors, only one of them is ever
switched on in a given olfactory sensory neuron,
enabling the cell to sense only a small variety
of scent molecules. Once they detect an odorant
molecule, these neurons send along their
findings to the olfactory bulb, the first stop in
the smell pathway.

     This image was obtained in an elegant
experiment in which a blue stain is switched on
in only the small fraction of olfactory sensory
neurons that use a particular receptor gene.
We see these neurons' somata and dendrites
(the blue specks to the left) lining the span of the
nose and projecting axons over to the olfactory
bulb (at right). A glance is sufficient to grasp the
logic of this particular circuit, a testament to how
molecular biology can inform us about how the
brain organizes information: Neurons that use
the same receptor gene span the entire lining
of the nose, but their axons coalesce and carry
information to one specific spot in the olfactory
bulb, where it is collected for further processing.

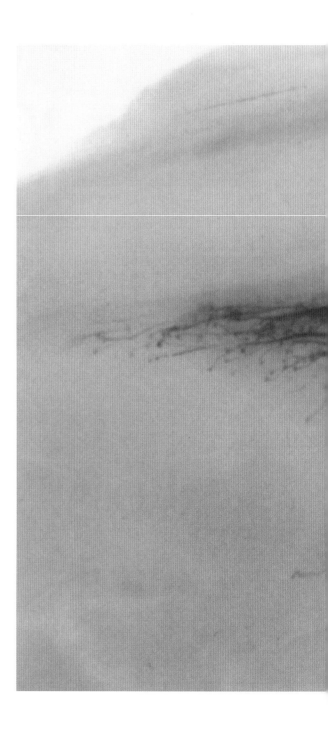

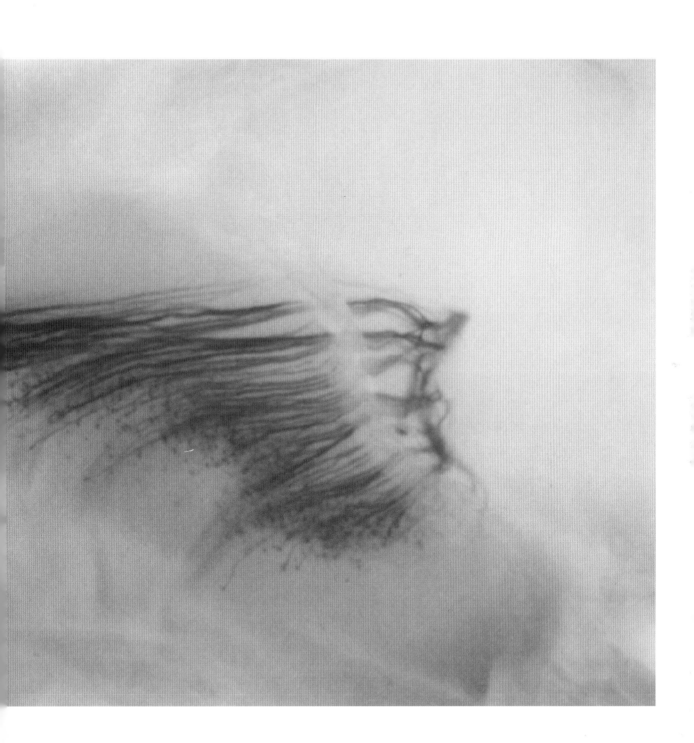

Chick retina.
Andy Fischer, 2008.

This image of a chick's retina reveals the three
basic stages of processing by the circuit that
captures light and translates it into signals
the brain can understand, thus enabling us to
see. At the top of the image are the retina's
photoreceptor cells (in gray)—the familiar rods
and cones—that capture photons of light and
translate them into electrical currents. These
rods (which detect dim light) and cones (which
enable color vision during daylight) pass on their
information to an intermediate layer of neurons
(in the middle, in red and orange) for further
processing. Unlike those in the nose, the neurons
in the retina perform a significant amount of
analysis on the information flowing in before
sending it further into the brain. In the bottom
layer (dark orange), retinal ganglion cells sum
up the retina's findings and translate them into
action potentials.

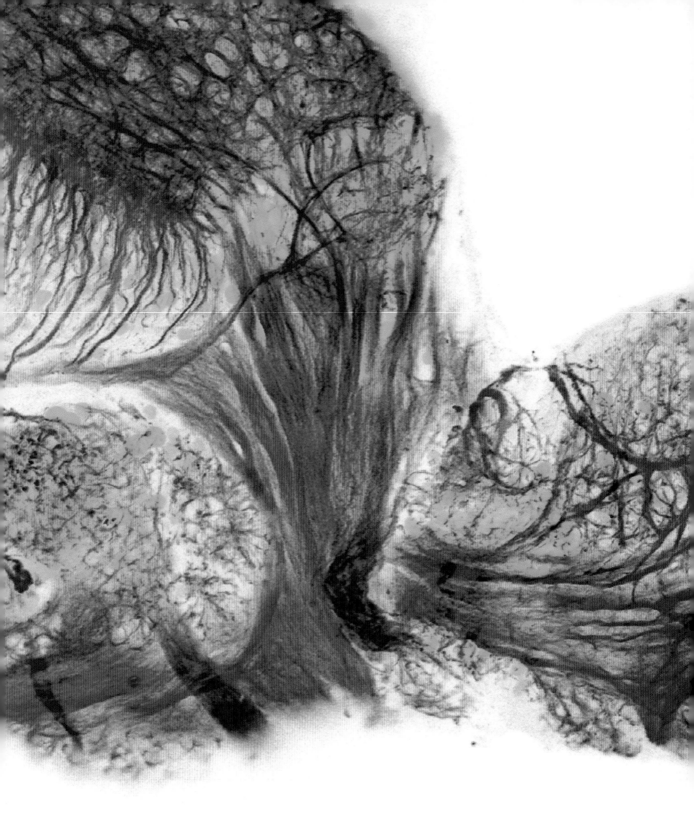

Zebrafish retinal axons.
Michael Hendricks and
Suresh Jesuthasan, 2006.

In this image, we see the axons that carry the
retina's output to the rest of the brain in the
wedge-shape pathway (middle) leading to the
large green area in the upper left. They are made
visible by an antibody staining that marks all
axons (pseudocolored in brown).

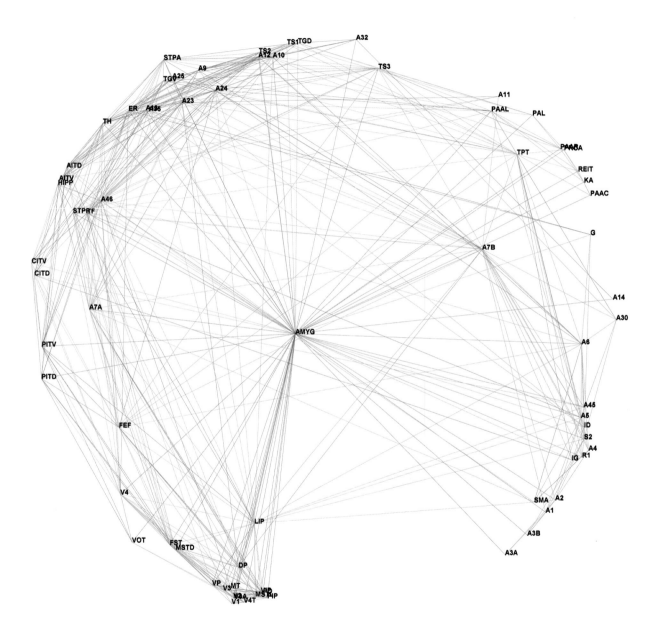

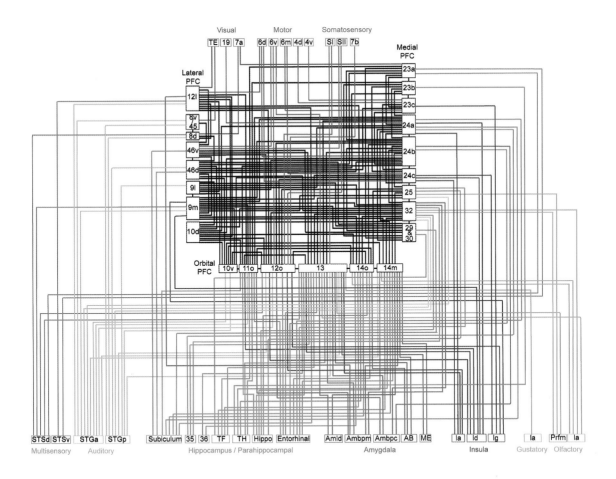

Opposite: Amygdala connectivity map.
Malcolm Young, Jack Scannell, Gully Burns,
and Colin Blakemore, 1994.
Above: Frontal cortex connectivity map.
Bruno B. Averbeck and Moonsang Seo, 2008.

These two images illustrate neural circuits at
an abstract level, as pure wiring diagrams
reminiscent of an engineer's electronic circuit.
These kinds of maps, derived by synthesizing
raw data obtained in the course of hundreds
of experiments, represent the known long-range
connections between regions of the brain and
are used to understand how a particular area
of interest relates to others.

The first image (opposite) is a summary
of the connections between the amygdala, an
area deep inside the brain that processes our
emotions, and the cerebral cortex. The amygdala
is represented here in the center of this constel-
lation; all the areas in the cortex, including many

of those responsible for reasoning and high-
level decision making, are shown surrounding
it. This diagram reveals that all but eight of
these areas are directly connected in some
way to the amygdala, anatomically linking
areas traditionally thought of as serving
"rational" or "emotional" features of the mind.
The many connections between these areas
argues for a view in which thought and emotion
are less dissociable from each other than is
commonly believed.

The image above focuses on the frontal
cortex, the area of the brain traditionally thought
to be responsible for higher mental functions
such as planning, deciding between good and
bad courses of action, and suppressing instinc-
tual or habitual behaviors. Again, massive interconnectivity is the organizing principle: Sensory,
emotional memory, and motor systems all feed
information to this area, making it a veritable hub.

Sensory Neurons   Interneurons   Motor Neurons

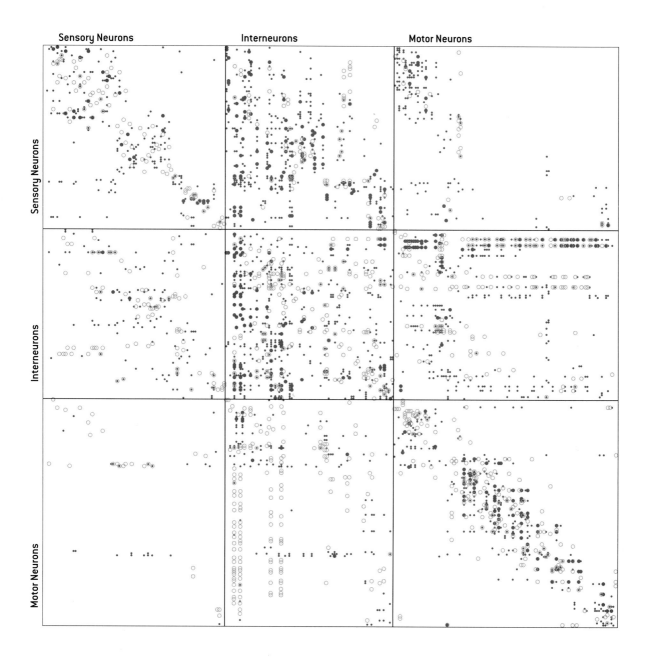

Caenorhabditis elegans connectome.
Lav Varshney, Beth Chen, Eric Paniagua,
David Hall, and Dmitri Chklovskii, 2009.

The preceding diagrams offer a broad view of brain circuitry—they describe how a group of neurons in area A project to a group of neurons in area B. This is critical information, but despite its great complexity, it still does not account for the richness in connectivity patterns in the brain. Each single neuron can connect to (and be connected by) thousands of different neurons, adding another significant layer of complexity to an already daunting picture. We must now ask: For a given neuron in area A, which specific neurons in area B does it connect to? This assumes, of course, that a selective connectivity pattern between neurons in A and neurons in B exists and is an important property, one necessary to the proper function of the system (and one that, if impaired, could give rise to dysfunction). We know that at least a moderate level of specificity is required. The question then is, do we need to probe even deeper to uncover the neuron-by-neuron connectivity—and if not, what level of precision is necessary to capture the essential features of the system? Unfortunately, this is a question that likely cannot be resolved in the abstract, no matter how brilliant the argument. Nothing short of a full account of neuron-by-neuron connectivity can reveal whether a diagram so precise is in fact necessary. There are strong voices on either side of this question, as a number of large-scale endeavors have been undertaken to uncover the brain's *connectome*—the definitive wiring diagram, much like a genome is the definitive sequence of genetic information—at various levels of detail.

The best way to commence this daunting task is to start small. *C. elegans* is a one-millimeter-long brainless worm, with a noted fondness for compost piles. Its entire nervous system contains exactly 302 neurons, making it ideally suited as a subject for generating a synapse-by-synapse wiring diagram. Begun decades ago, the *C. elegans* connectome project recently yielded a nearly complete account of every connection between these neurons—the very first diagram at such a fine scale of any animal's nervous system. It was created by painstakingly following the extensions of each neuron through consecutive, infinitesimally thin slices of the worm, each examined in an electron microscope. The diagram shown here divides the neurons into three classes: *sensory neurons*, which detect features in the environment; *motor neurons*, which output commands to muscles in order to generate motion; and *interneurons*, those in between. Each individual neuron is plotted on both the vertical and horizontal axes, with a dot representing one of the thousands of connections found—red for chemical synapses involving neurotransmitters that pass information from one neuron to the next; and blue for gap junctions, in which two neurons are directly connected to each other through pores in their membranes. One glance at this graph is enough to detect salient properties about the circuit: For instance, sensory neurons and motor neurons do not tend to be connected (bottom-left and top-right squares), whereas interneurons are quite densely interconnected (central square).

A sophisticated—and, who knows, perhaps even complete—understanding of how *C. elegans*'s nervous system relates to its behavior is far in the future. This will require many more experiments, whose design and interpretation are now made possible by this complete map of this organism's circuitry. Even though the structure of a neural network can yield tremendous insight into how it governs function and behavior (a principle that Cajal exploited with great success), a map alone cannot tell the entire story. The vast project of examining the functional contributions of each one of its parts is still in progress.

Automated reconstruction of serial block-face scanning electron microscopy data.
Viren Jain, Joseph Murray, Sebastian Seung, Srinivas Turaga, Kevin Briggman, Winfried Denk, and Moritz Helmstaedter, 2007.

Mammalian brains contain vastly more neurons than *C. elegans*'s convenient 302. While it has proved possible (heroic, actually, considering the tediousness of the task) to generate a complete wiring diagram for the worm, the approach employed is simply not going to work for larger nervous systems like those of mice or—especially—in the human brain, which contains an estimated hundred billion neurons. Very recently developed tools like Brainbow (page 91) and serial block-face scanning electron microscopy (page 131) address this exact problem by separating vast amounts of information by color and automating the collection of high-resolution data, respectively. But both techniques are limited, paradoxically, by the dizzying volumes of data they generate, which must then somehow be interpreted in order to pinpoint where each neuron makes synaptic connections with others. Estimates of how long it would take to trace circuits in nervous systems more advanced than that of *C. elegans* run up to tens of thousands of years of meticulous human study. Graduate students can be driven hard at times, but there are limits.

The only scalable solution is to program computers to do the "seeing" and the tracing of neurons for us, in order to determine where they make synapses. In recent years, experimental neuroscientists have increasingly welcomed into their ranks specialists from other more theoretical traditions (physics, computer science, and mathematics) to work together and devise ways of making sense of the raw data recorded in the course of experiments. These theoreticians have quickly become involved in many areas of the field, including connectome projects. Without their partnership, the goal of determining a mammalian nervous system's connectome at the level of detail now available for *C. elegans* would simply prove impossible. The image on the opposite page shows neurons that were algorithmically traced through a series of hundreds of consecutive slices of a rabbit's retina, which had been imaged automatically with a serial block-face scanning electron microscope.

It is worth pointing out that the original fire that fueled a century of neuroscience—methods (like the Golgi stain) that strip away large amounts of distracting information in order to focus only on a few structures of choice—is now being passed over by some in favor of gathering all available details in all their confusing glory. The hope is that we can somehow make sense of them through computer-assisted analysis. This is truly uncharted territory and a radical shift away from the approach so successfully pursued for a century by Cajal and his heirs.

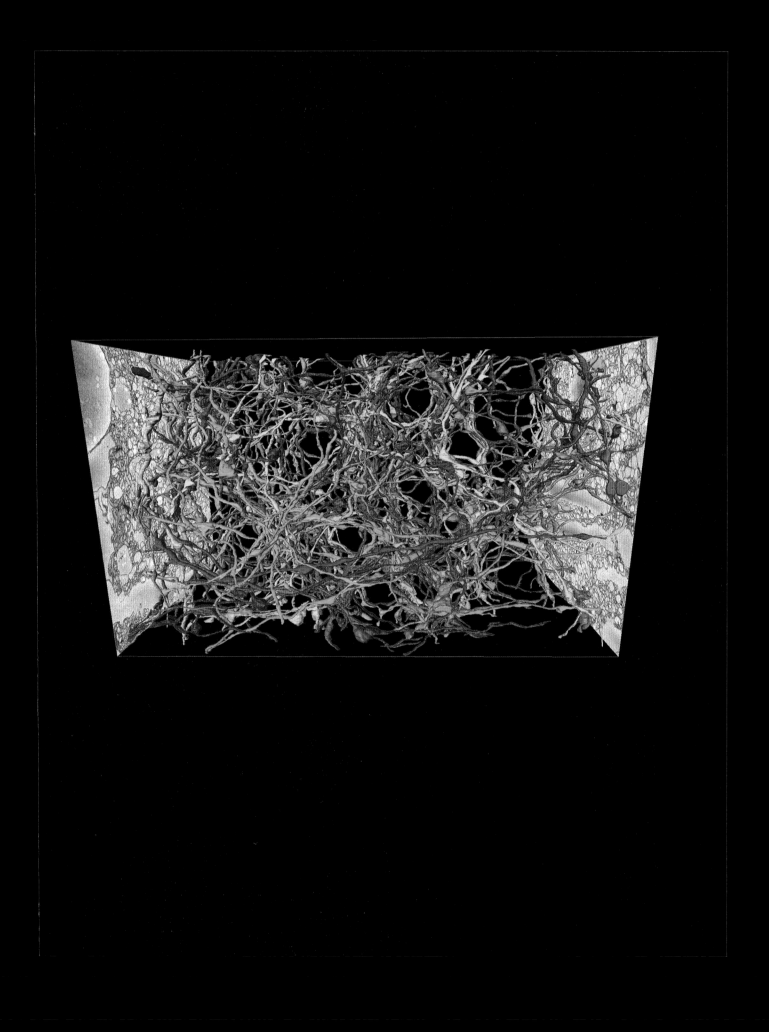

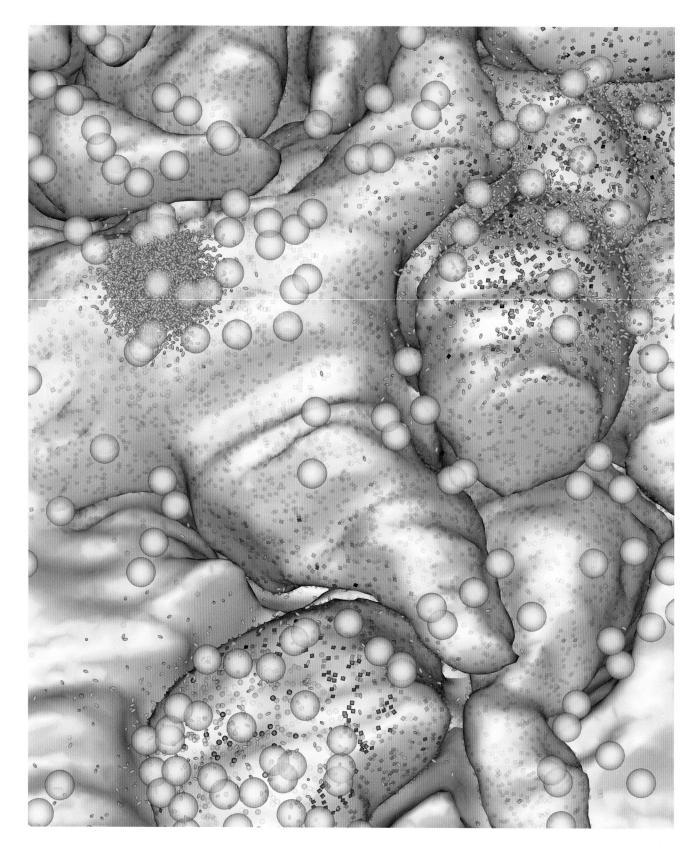

MCell.
Tom Bartol, Jay Coggan, Eduardo Esquenazi,
and Terrence Sejnowski, 2005.

Computer-assisted analysis enables theoretical neuroscientists to extract meaning from the avalanche of data pouring in from all sides and to make predictions that can then be used by experimentalists to move forward in our understanding of the brain. The field's experimental insight is hobbled by a devilish combination of vastness and complexity, which renders impossible any attempt to, for instance, record the activity of every neuron in even a simple network—or, even more challenging, the state of its every synapse. The phenomena that occur in the brain are extraordinarily fast, difficult to capture even with today's technology, and cannot, as of yet, be measured on a large scale. One of the critical tasks for theoreticians, then, is to take all of what we can access and generate computer simulations of the larger processes, of which we can experimentally observe only the very tip of the iceberg. At their best, their simulations suggest broad principles that may emerge from the chaos underneath, which can guide experimental research.

Obtained using custom-made software called MCell, this image is a snapshot from a very-fine-scale computer simulation of the diffusion of individual acetylcholine neurotransmitter molecules at a synapse on a microsecond scale. It is based on a high-resolution anatomical reconstruction of a synapse in the ciliary ganglion of a chick, which was imaged with an electron microscope. Molecules of the neurotransmitter (green ellipsoids) are released from synaptic vesicles (yellow spheres) of one neuron and diffuse across the synaptic space between the two neurons. The neurotransmitter must then bind to receptor molecules in the second neuron's membrane (blue and red objects) before being destroyed by enzymes nearby in the synaptic space (not shown). If enough receptors are activated, this sets off a cascade of events that results in the activation of the second neuron. Because this kind of simulation can model synaptic transmission on a molecule-by-molecule basis, it can generate precise quantitative predictions, some of which have already been confirmed experimentally.

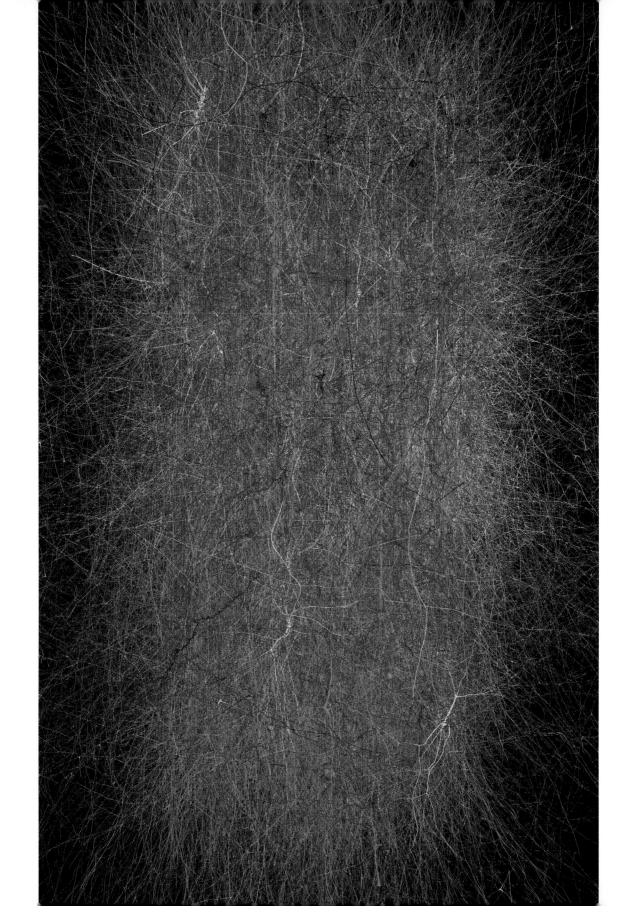

Blue Brain Project.
Henry Markram, 2008.

This image shows the three-dimensional
configuration of ten thousand simulated neurons
that constitute a single *neocortical column*—
an anatomical unit barely wider than the head
of a pin. The neocortical column, believed to be
a building block of the cerebral cortex, is a mere
millimeter cubed, and is repeated countless
times across the expanse of the human
neocortex. Since there is simply no way to gain
full experimental insight into a process at this
large a scale, the Blue Brain Project was launched
in 2005 in collaboration with IBM to produce a
computer simulation of it. It is so spectacularly
complex that a dedicated state-of-the-art
supercomputer is required to keep track of all
the phenomena as they recurrently influence
one another; even with this computational
firepower, it still takes about one hundred
seconds to simulate a single second of activity.
In order to make it as realistic as possible, a wide
variety of factors is included in the simulation:
genetics, the shape of dendrites, and the
neurophysiological characteristics imparted
by their ion channel composition. Here we see
the Blue Brain's cortical column in action: Each
dendrite is simulated individually and rendered
so that its color (ranging from blue to red)
represents its voltage at one moment in time.
The long-term goal of the project is to uncover
the broad principles of brain function and
dysfunction by simulating the entire brain of
mammals, including that of humans.

Simulating epilepsy.
Larry Abbott, 2009.

Instead of incorporating all of the physical details into a simulation, an alternate and complementary approach to gaining insight into the principles that govern neural networks is to abstract the brain to a minimal set of fundamental elements. This image represents a simulation of a network of one thousand neurons, each "connected" to 10 percent of the others; it shows the activity of only one hundred of these simulated neurons (one to each row), with each notch symbolizing an action potential, the electrical unit of information in the brain. The simulation displays epileptic properties in which all the neurons in the network synchronously emit a barrage of spikes every second or so. (An actual patient would exhibit typical seizure symptoms during these volleys.) The network then reverts back to normal activity, as shown in the wide column at the far right. Computer simulations such as this one allow theoretical scientists to investigate the key mathematical properties of networks that give rise to pathologies like seizures.

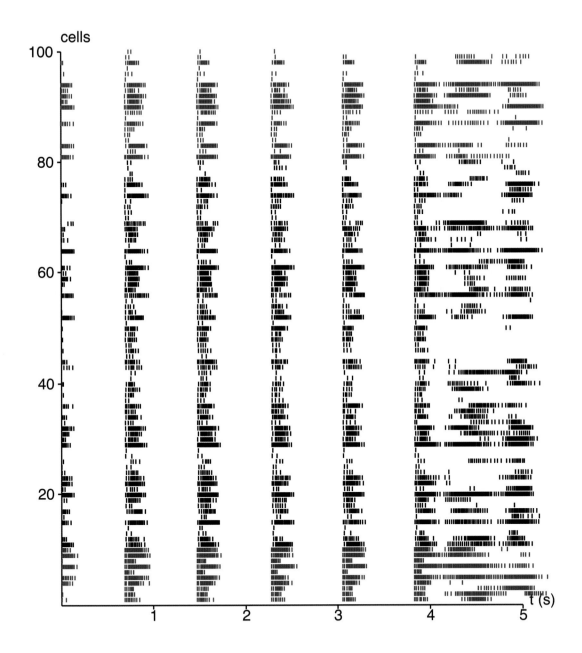

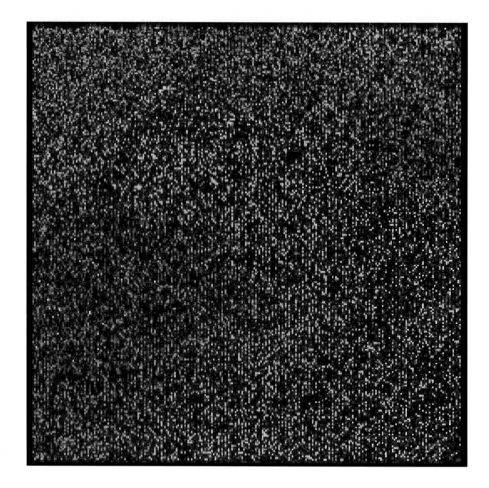

BRAIN ON A CHIP

Above: Chip dynamics.
Opposite: Chip design.
John Arthur, Rodrigo Alvarez,
and Kwabena Boahen, 2008.

This chapter closes, fittingly, with a look at
an actual computer circuit, a "brain on a chip,"
designed to simulate a large network of
neurons. Unlike the chips in a computer, this
one is designed to be brainlike. The passage
of ions through channels in a neuron's
membrane is mimicked by the flow of electrons
through transistors on the silicon (opposite).
This method saves staggering amounts of time,
electrical power, and money, since scientists
can simulate neurons in real time directly
on hardware, instead of using software to
painstakingly calculate their dynamics using

complicated equations. The image above is
an example of what can be obtained from such
chips: It is a simulation of a population of 65,536
neurons (each one represented by a square)
showing that adjacent neurons in the network
can be locally in sync and at the same time be
largely out of sync with the rest of the network.
(In this simulation, neurons that are in sync
share the same color.) As it develops, the
technology promises to free theorists from the
need for expensive supercomputers to crunch
dizzying numbers—and from the equally
crushing task of having to apply for enormous
grants to buy and run these computers. What
today requires a state-of-the-art dedicated
facility might one day in the future be available
to enterprising teenagers.

In light of this and other large-scale projects, it is worth reflecting on the role, in scientific understanding, of modeling natural phenomena; surely some readers will object that building a large computer simulation is necessarily inferior to physically probing an actual brain. Yet when we say that the Earth rotates around the sun, we do not think of the countless atoms that constitute the Earth, moving in very precise directions and at specific speeds around the sun's own set of countless atoms. While the latter description of the process is undeniably more precise, the first, in most cases, is more useful: A model is valuable when it collapses an unmanageable amount of information into a simplified set of principles that captures just enough about the system to provide insight into how it behaves. Thus, the statement "The Earth rotates around the sun" can be considered a tacit model, and in this sense much of our scientific knowledge is in model form. An understanding of the brain, the most complex and unmanageable organ we have yet encountered, requires the development of models. In the words of a theorist: "In the brain the model is the goal."[1]

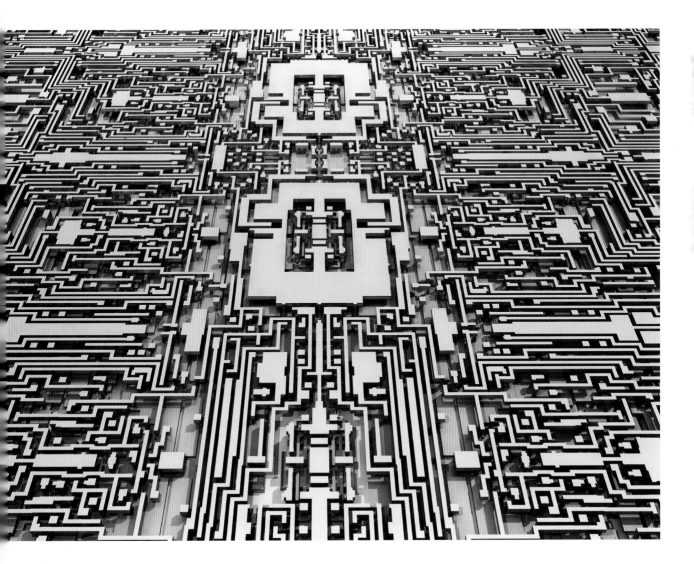

7.

FROM BRAIN STRUCTURE
TO BRAIN FUNCTION

FROM BRAIN STRUCTURE TO BRAIN FUNCTION

by Joy Hirsch
Columbia University

During a recent meeting, one of my students confessed to being anxious about not knowing what the outcome of her functional neural imaging experiment would be. I smiled because this anxiety is, I think, a hallmark of good, interesting science. This kind of uncertainty arises often in the context of contemporary neuroimaging, which seeks to uncover the principles that relate brain and behavior by imaging its activity in a live subject. I shared with her my basic rule of thumb: "In science, the unknown is our home base, and if you know what you're doing, you shouldn't be doing it. We can rely on the data for answers without prior knowledge." This is of course an oversimplification, but my student got the gist of it and reinterpreted her hesitation about the unknown as the thrilling experience of advancing our nascent knowledge about mind and brain.

Although a relatively new discipline, functional neural imaging has already taken us far into the unknown territory of the biology that underlies who we are, altering how we think about ourselves and each other. In *The Astonishing Hypothesis*, Francis Crick eloquently illuminates this connection: "You, your joys and your sorrows, your memories and your ambitions, your sense of personal identity and free will, are in fact no more than the behavior of a vast assembly of nerve cells and their associated molecules."[1] The power of this "astonishing hypothesis"[2] is potentially on par with his earlier discovery of the molecular structure of DNA and the genetic code that builds all living things. Although our knowledge of the basic mechanisms linking biology and behavior remains fuzzy, we can point to strong evidence for the intimate relationship between mind and brain in findings such as Erik Kandel's breakthroughs concerning the molecular basis of memory and Richard Axel's discovery of the receptors that endow us with our sense of smell. Widespread and growing acceptance of the idea that the mind is the brain—that we can understand our mental life in terms of our biological understanding of the brain—is evident in the neurocentric vocabulary that has recently begun to enrich our everyday language—words such as neuromarketing, neuroeconomics, neuropolitics, neurophilosophy, neuroethics, and neurolaw. As these hybrid words suggest, traditionally segregated academic disciplines are beginning to coalesce under the mantle of a "life sciences" program with the neuron as its focus. Under this view we can begin to unify the traditions of law, business, finance, economics, marketing, and psychology—not to mention the medical specialties of neurology, psychiatry, surgery, medicine, and radiology. New fields such as "decision sciences" have quickly gained ground, promising to build upon our knowledge of neural circuits in order to enhance our understanding of longstanding themes such as personal identity, free will, and the biases we exhibit when faced with uncertainty or risk.

Within ourselves, this new neurocentric view has many advantages. I now have new tools to cope with everyday anxieties such as when I board a plane or get stuck in an elevator—"Amygdala, be still," I sometimes catch myself intoning. My motivation to resist a chocolate chip cookie gains strength when I recall a neural model of cognitive control that translates this kind of temptation into a "tug-of-war" between the neural systems that govern reward and executive control. And so I "tug" for the executive control team and strengthen my resistance against the chocolate chip cookie. When I answer incessant phone calls, read the daily deluge of e-mail, or navigate the treacherous political landscape of academic life, my "vast assembly of nerve cells and their assorted molecules"[3] (Crick's formulation) fatigues with overuse. By eating, sleeping and relaxing, I can nurture these neurons and keep my personal forces intact.

These cells and molecules, awash in various neurochemical cocktails in my basal ganglia, are presumably the basis for my love and attachment to my husband. Earlier in my academic journey I would have resisted this unavoidable fact of biology on the misguided grounds that a physical basis would diminish the grandeur and centrality of my choice of a life partner. This dismissive (but familiar) view of our biology is shared by the Cheshire Cat in *Alice in Wonderland*, who shrugs off Alice's distress and pronounces, "You're nothing but a pack of cards." (Or, in Crick's retelling of it: "You're nothing but a pack of neurons."[4]) Yet my experience as a neuroscientist reveals facets of how "[my] joys and [my] sorrows and [my] memories and [my] ambitions, [my] sense of personal identity and free will,"[5] arise, and helps me make better sense of my experiences. This neural science is an heir to the research programs that have enabled humans, over the centuries, to better understand who we are and how we fit into the universe.

＊　＊　＊

We have long been aware that human experience begins with the outside world bombarding us with packets of energy. For instance, neurons in a brain area called Heschl's gyrus transduce the energy of air pressure detected by delicate membranes in the ear into the perception of sound, giving rise to our experience of language, music, meaningful information, and their associated emotions. Specialized neurons in the back of the eye convert energy from photons and a specific range of radio frequencies into the perception of light; milliseconds later we are awash in colors, shapes, objects, and meaningful visual scenes that connect us to the world and to each other. The smell of a cup of Starbucks coffee begins with an airborne molecule that is captured by dedicated neurons in the epithelial layer of nasal passages and converted into the delightful aroma that I so love in the morning.

If neurons are at the root of who and what we are, how do they do it? The history of science is filled with examples of how quantum leaps in our understanding were made possible by advances in imaging technology. The microscope enabled views of the infinitely small and the telescope enabled views of the infinitely large, each extending the boundaries of the human eye and with it the boundaries of science, leading to a deeper understanding of the principles that govern the universe. The current imaging technology in mainstream neuroscience is equally revolutionary, for it allows us to simultaneously view a brain's structure and its function, enabling us to investigate the relationships between them. I think of this dual-imaging research strategy as a "mindoscope"—a tool that can probe the workings of the mind (the intangibles of conscious experience) and establish how they are linked to the anatomy and function of the brain (structure, chemistry, and neural activity). Under this research program, one first reveals the precise shape of a living human brain using conventional magnetic resonance imaging (MRI; see pages 206–209), and then highlights in this image the specific neural circuits that are engaged by a given behavioral task detected by functional magnetic resonance imaging (fMRI; see pages 224–227). This detection is made possible by the fact that when an area of the brain dedicated to a specific task—like object naming—is activated, its consumption of metabolites carried by the blood increases. fMRI detects the changes in the magnetic susceptibility of oxygen-saturated blood recruited by these active neurons, and we interpret these changes in magnetic properties as a proxy for neural activity. To be clear: We do not claim to measure brain activity directly, but we follow the basic rule that more blood flows into a particular brain area presumably because that area is more active. Different levels of activity are then translated to colored blobs superimposed upon the picture of the brain—the core data of neural-imaging studies. The biology that underlies the roots of conscious experience, including perception, thought, and action, is revealed by experiments that can selectively engage the specific neural circuits associated with these faculties.

As with any tool in science, fMRI and complementary imaging techniques such as PET, EEG, and optical imaging are subject to limitations. Some of the hardest questions—How do neurons transform energy packets captured from the environment into conscious experience?—still remain far out of our reach. Nonetheless, current mindoscope researchers can segment the healthy brain according to task-related systems. When these studies employ rigorous experimental design, they reveal dynamic mechanisms of how communication pathways in the brain can exert influences upon one another and so underlie our experiences and behaviors. Some notable recent discoveries from fMRI research labs include mechanisms that suppress emotional responses, boost detection processes, and improve focus on cognitive tasks.

✳ ✳ ✳

PORTRAITS OF THE MIND

With the advent of functional brain imaging studies, a floodgate of new opportunities to advance how we think about our behavior and ourselves has opened up. This science will ultimately translate into benefits for medicine and society. Our ability to measure the biological mechanisms of human behavior using brain imaging introduces a new dimension of objectivity into our investigations, which will help translate research findings to advances in patient care and quality of life. A few recent examples stemming from my research program follow, but there are many others. Eating disorders, obesity, and addictions can be studied in terms of the neural circuitry sensitive to reward and executive control. Anxiety disorders and post-traumatic stress may someday be diagnosed, prevented, and treated by image-guided approaches that reveal the circuitry related to fear. In the future, developmental disorders such as autism may be diagnosed and treated using strategies based on methods guided by functional imaging. Psychiatric conditions such as schizophrenia, as well as obsessive-compulsive and panic disorders, may someday be diagnosed by characterizing a patient's neural circuitry; this analysis would then inform customized treatment strategies. A recent finding suggests that patients with traumatic brain injury leading to disorders of consciousness may one day be "given a voice" by imaging technology.

The possible gains extend beyond the frontiers of medicine. Neural systems engaged during economic decision-making processes in healthy individuals mediate biases related to fear, emotional arousal, and recent losses and gains, which may alert financial managers to "danger situations" related to financial decisions. Value and preference may be linked to neural underpinnings, as well as political attitudes, prejudices, and moral judgments. Learning may be guided by neural tendencies, and compensatory mechanisms for memory loss may be neurally guided and strategically enhanced. Even national policies may one day be studied in terms of neural responses related to qualities such as altruism, personal comfort, and perceived safety. The transformative effects of art, poetry, and music could also be understood in terms of gifted neurons endowed with complex abilities to connect highly valued stimuli to neural emotion-generators in the brain.

As these speculative examples illustrate, a vast array of questions, insights, and medical advances are about to emerge from teams of scientists, physicians, and students that will undoubtedly impact our understanding of the human condition. In the context of this new framework based on the astonishing unity of the physical brain and the mind, we all become students and together take the neuroscience journey along the road to the mind. This journey affords us an unprecedented opportunity to understand ourselves as glorious works of nature. Under this view, our fertile neurons not only solve our day-to-day problems, they develop our very experience of the joys and sorrows, the love and beauty of life itself.

Positron emission tomography imaging.
Mark Slifstein and Anissa Abi-Dargham, 2010.

Because of the complexity of the nervous
system, and our still-poor understanding of it,
neuroscience has been far slower than many
other fields of biomedical research to translate
its findings into clinical applications. However,
one area that has provided many tangible results
to patients (not to mention invaluable insights
to researchers) is live brain imaging.

      Positron emission tomography (PET) can
be used to track specific molecules, such as
neurotransmitter receptors, in the brains of
live subjects. The data was obtained by injecting
a schizophrenia patient with a radioactive
tracer that binds dopamine receptors, allowing
detection by a PET scanner. In the study shown
here, researchers imaged these receptors before
and after temporarily depleting the subject's
dopamine with a chemical agent. By comparing
the before and after images (top and bottom
respectively), they could quantify how much
dopamine schizophrenia patients have in a brain
region called the striatum. This measurement
was also performed in healthy subjects, and
the comparison revealed that schizophrenia
patients exhibit twice as much dopamine in this
area, suggesting targets for future therapeutic
strategies. (Note that the bottom image is
brighter, implying that there were more receptors
available to bind the radioactive tracer, and
therefore less remaining dopamine.)

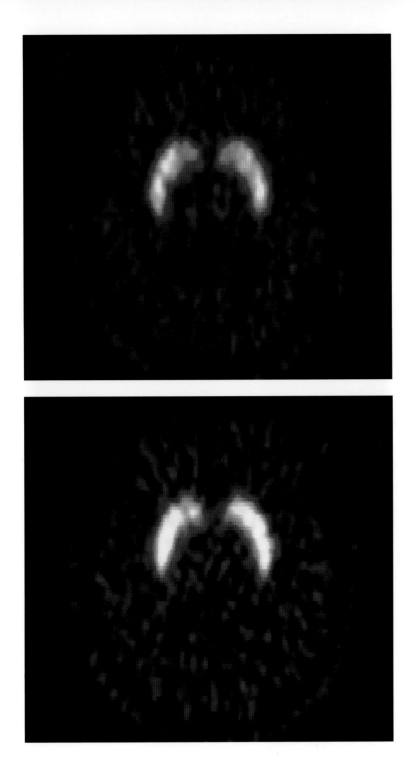

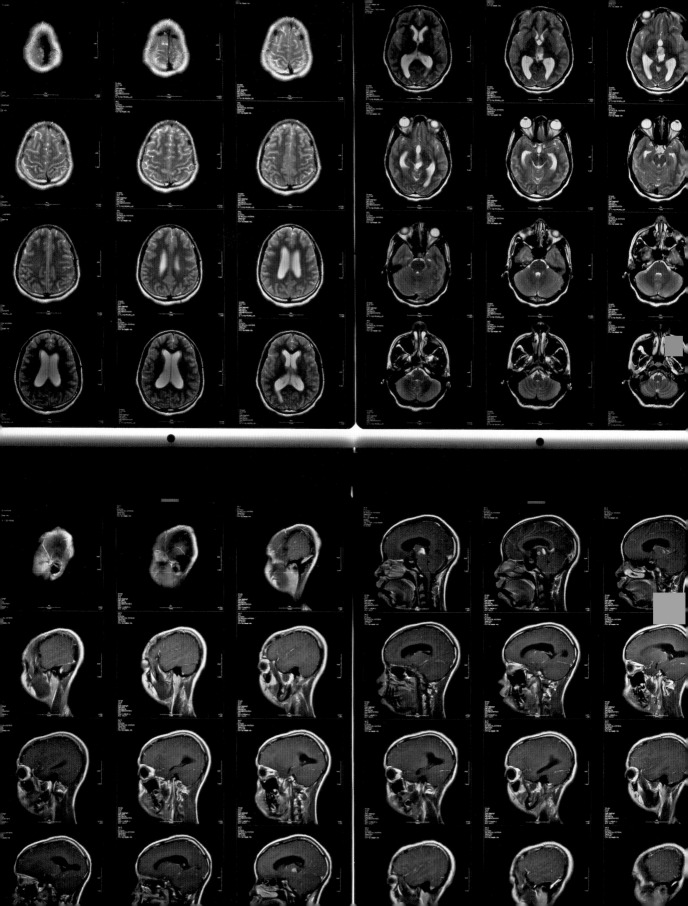

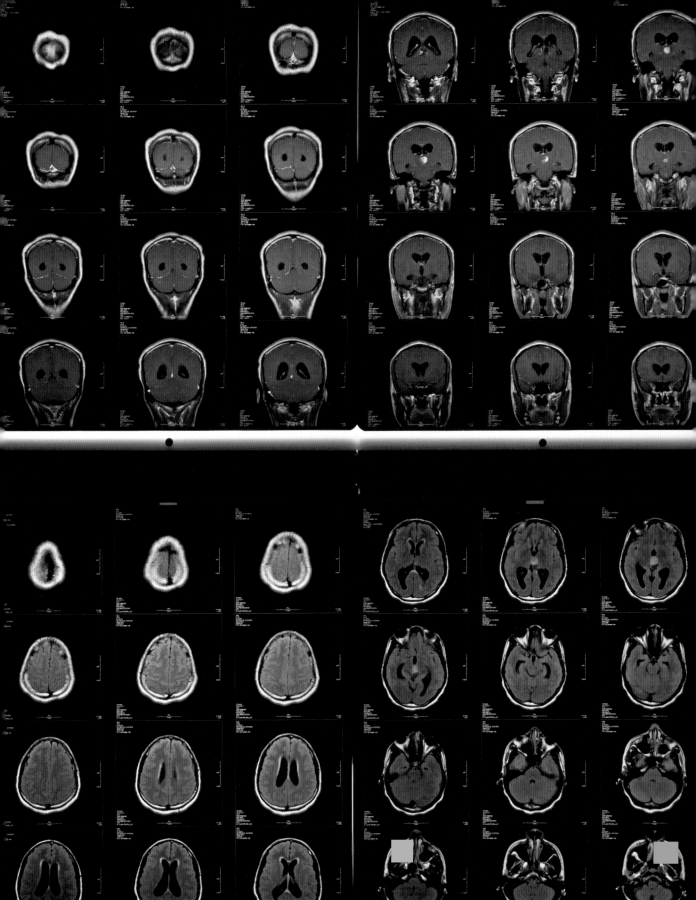

ANATOMICAL MAGNETIC RESONANCE IMAGING

Pevious spread: Overview.
Opposite: Detail.
Raqeeb Haque, 2009.

Magnetic resonance imaging, or MRI, is a widespread tool in both research and clinical contexts. Although MRI does not afford the same level of resolution as a microscope, it has the distinct advantage of being noninvasive and harmless. An MRI machine is essentially a gigantic magnet—and the bigger the better. (Ubiquitous in MRI labs are signs warning visitors against carrying anything metallic, lest it should fly out of their pockets and crash into the apparatus. A host of spectacular safety videos available on YouTube illustrate the consequences of ignoring this injunction; one particularly chilling one involves an oxygen bottle and a watermelon.)

The subject is placed within the magnetic field, which aligns the hydrogen protons in the body's water molecules like countless compasses, all pointing in the same direction. The machine then administers short radio wave pulses, which cause the hydrogen protons to become momentarily nudged into a different alignment. Very quickly thereafter they relax back to the original alignment, but since different types of brain tissue cause different relaxation rates, these can be measured and used to identify white matter (containing axons) from gray matter (somata and dendrites). During an MRI procedure, the subject experiences a deafening sound—dampened as best as possible with earplugs—and the irritation of having to lie perfectly still for a long stretch of time in a narrow, enclosed space.

Even though MRI cannot detect individual neurons, it provides invaluable information about the shape of an organ as a whole. These two images show sequences of virtual "slices" through a patient's brain. Using images like these, physicians can identify malformations and injuries without having to open up the skull and examine the brain directly. The patient shown here was found to have a tumor in the thalamus, as well as an excess accumulation of fluid inside the brain. Using the anatomical data shown here, surgeons were able to surgically remove the tumor and drain the fluid.

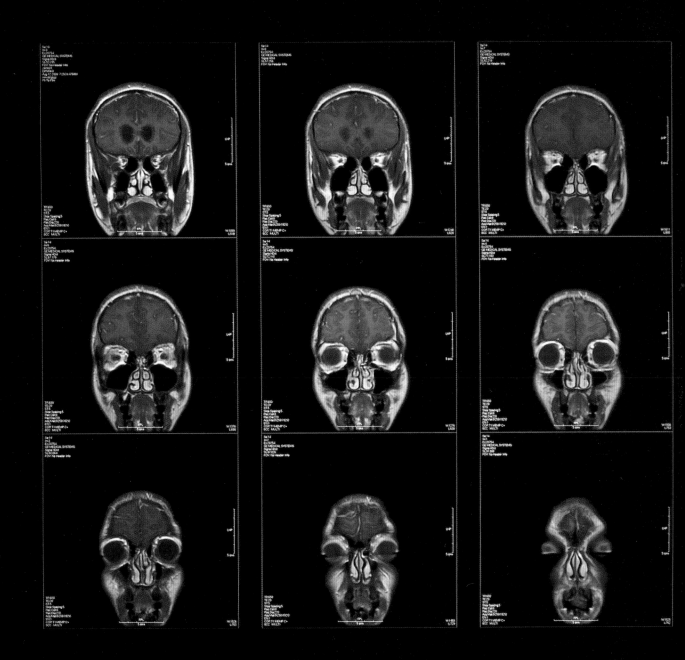

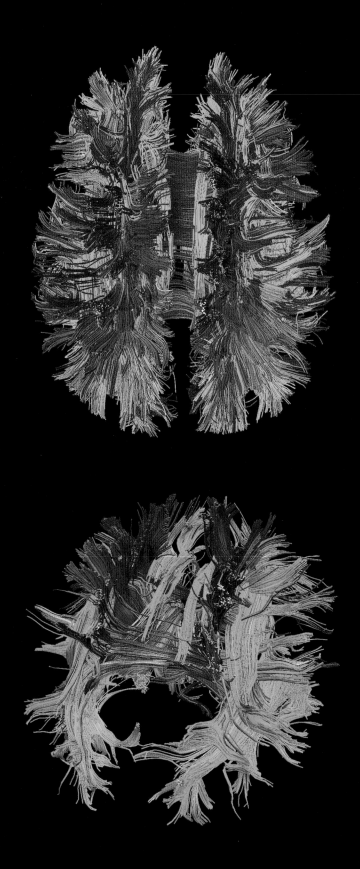

Diffusion MRI.
Patric Hagmann, 2006.

MRI machines are remarkably versatile, and
new uses for them are constantly being
discovered. These two images show how MRI
can be employed to establish the brain's major
anatomical features, but the technique isn't
limited only to large structures. A wide variety
of scanning modes enable researchers and
clinicians to detect an array of information, such
as individual pathways connecting separate
areas of the brain and even brain activity. A new
method called diffusion MRI can be used to
uncover major axon pathways in the brain by
measuring the motion of water contained within
a group (or tract) of axons traveling from one
point in the organ to another. This method is
capable of detecting the natural diffusion of
water molecules along these tracts—again,
completely noninvasively—and thereby infer
their paths indirectly.

The two images opposite show a tracto-
graphy of a human brain, obtained in a live
human subject who walked out of the apparatus
unharmed. At top, we are looking down on the
brain, with the back of the head at the bottom
of the image and the forehead at top; in the
view below it, we are looking at the subject
from the back of the head. Each line does not
represent a single axon, but thousands of them,
traveling together as a group. The colors indicate
the axis of each fiber (green: front to back; red:
left to right; blue: top to bottom).

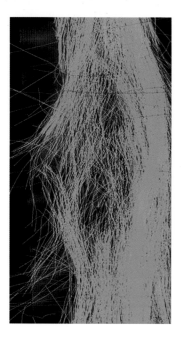

DIFFUSION MRI:
CLINICAL APPLICATIONS

Above: Damaged spinal cord.
Denis Ducreux, 2007.
Opposite: Damaged thalamus.
Henning U. Voss and Nicholas D. Schiff, 2008.

Although still relatively new and still in development, diffusion MRI has produced much excitement in the field. These two images illustrate its power and promise as a diagnostic tool in clinical settings. The first (above) shows axons in a patient with abnormalities in the vasculature of the spine. As a result of this condition, insufficient oxygen is delivered to neurons surrounding the malformation, causing symptoms that can range from numbness and weakness to more severe conditions like paralysis.

The second (right) is from a patient who has suffered a stroke in the thalamus and midbrain, resulting in major disruptions to certain axon tracts, some of them visible at the bottom of the figure. With the resolution of these tractographies improving rapidly, diffusion MRI may someday become one of the standard tools for planning surgical procedures such as the removal of a tumor, as it can indicate to doctors which axon-rich areas must be avoided.

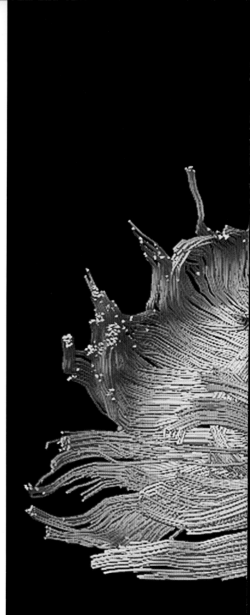

Diffusion MRI: research applications.
Patric Hagmann, Leila Cammoun, Xavier
Gigandet, Reto Meuli, Christopher Honey,
Van Wedeen, and Olaf Sporns, 2008.

On the research side, diffusion MRI has been
used in a wide range of studies, including one
in which axon abnormalities were found to be
linked to reading disability in children. Diffusion
MRI is at the center of an ambitious new initiative
to generate a detailed map of the entire human
adult brain's connectivity by 2015. Because of
the brain's sheer complexity—and the limited
means up until recently for studying it—our
knowledge of its pathways and hubs is still
incomplete. With the advent of diffusion MRI, and
help from its cousin, fMRI, the elusive goal of a
complete human brain map is now within reach.

This figure is only a taste of things to come: It
shows a broad map of the human cerebral cortex
connectivity patterns, as viewed from above (left)
and from the side (right). The size of each circle
indicates how strongly the area underneath it is
connected with others.

It is important to distinguish this macro-
scopic endeavor, which focuses on projections
between large groups of neurons, from parallel
tools like Brainbow, which promises to uncover the
connectivity between individual neurons at a
microscopic scale. The approaches are complementary—think maps of major interstate highways versus maps of city streets—and the hope is
that someday they will converge and produce a
complete account of the brain's connectome.

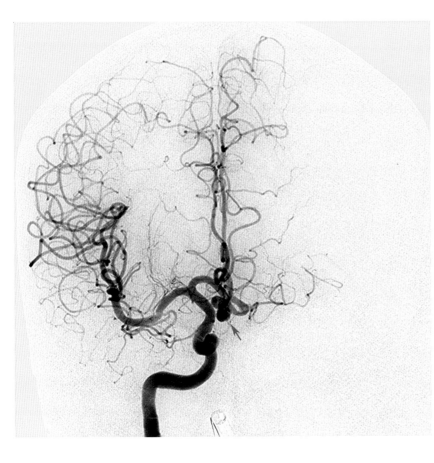

Cerebral angiography.
Kieran Murphy, 2009.

The images in this chapter so far have focused on nervous tissue, which leaves out a critical component of the brain's anatomy: its blood vessels. Even though the organ in our heads accounts for a mere 2 percent of the body's weight, it consumes one fifth of its oxygen at rest. An exquisitely elaborate network of arteries and veins circulates this precious commodity through the brain's every nook and cranny. Problems with blood vessels—such as malformations and aneurysms—can spell disaster, sometimes causing a debilitating loss of faculties such as speech and locomotion. In some cases, vascular problems can even be fatal.

Fortunately, an elegant and only moderately invasive method, first developed in the 1920s, provides the means to examine the brain's arteries. To obtain a cerebral angiogram, a physician guides a catheter to the carotid artery (a part of which is visible at bottom of the image), which supplies blood to the head, and injects a contrast agent that absorbs X-rays. An X-ray image taken immediately after an injection reveals the shape of the blood vessels, while all the rest—brain, bone, ventricles—barely register in the image captured on film. The method echoes Thomas Willis and Christopher Wren's early attempts in the seventeenth century to track the blood vessels of the brain with dyed solutions (see page 36). In the example shown here, the patient exhibited an aneurysm of the anterior communicating artery (arrow), which was treated by placing coils inside of it, causing the blood to clot, thus safely excluding it from circulation.

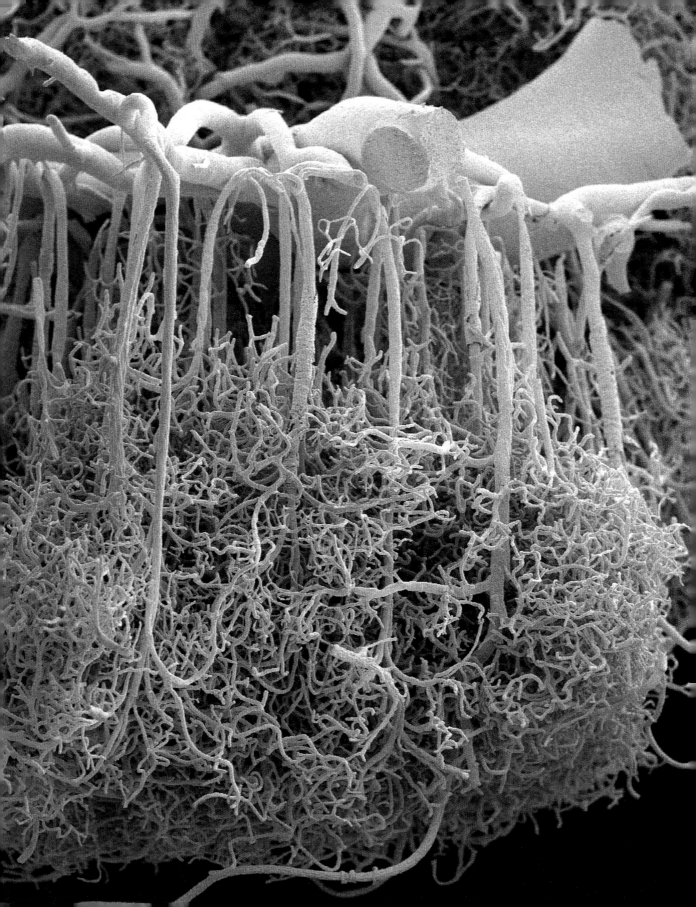

MICROVASCULATURE

Opposite: Human cerebral cortex.
Alfonso Rodríguez-Baeza and Marisa
Ortega-Sánchez, 2009.
Overleaf: Whole mouse brain.
Yoonsuck Choe, Louise C. Abbott, John
Keyser, Jaerock Kwon, David Mayerich,
and Bruce H. McCormick, 2008.

While cerebral angiograms provide a broad view
of the vasculature in a living subject, they cannot
detect the dense meshwork of microscopic blood
vessels that interface with cells in the brain.
Since these capillaries play a critical role
in fMRI studies, it is worth pausing to zoom in
(in postmortem samples) on the rich mesh that
furnishes the brain with blood. The image on the
following spread shows a mouse brain whose
blood vessels have been injected with India
ink to fill and stain them. The brain was then
automatically sliced and simultaneously imaged
using a custom-designed microscope. This image,
in which the blood vessels are represented in
white for clarity, is a view from the front of a brain
that has been sectioned through the middle.

The image opposite zooms in even further,
now in a human brain, with the aid of a scanning
electron microscope, to reveal the baroque
branching structures responsible for delivering
blood to the cortex. The organizational principle
is clearly illustrated here: Large blood vessels
surround the surface of the brain (at the top)
and send thin, dense projections down into the
depths of the cortex (below).

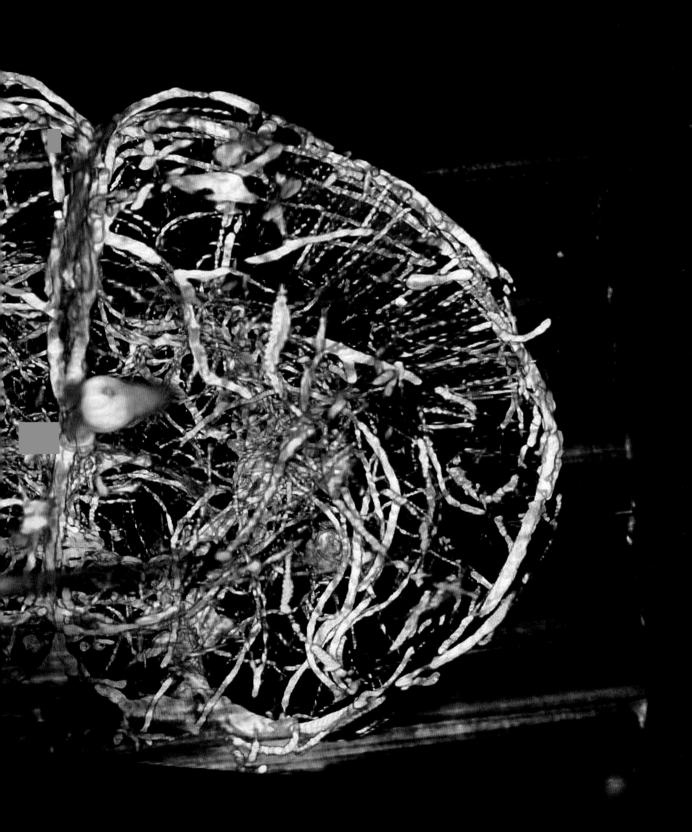

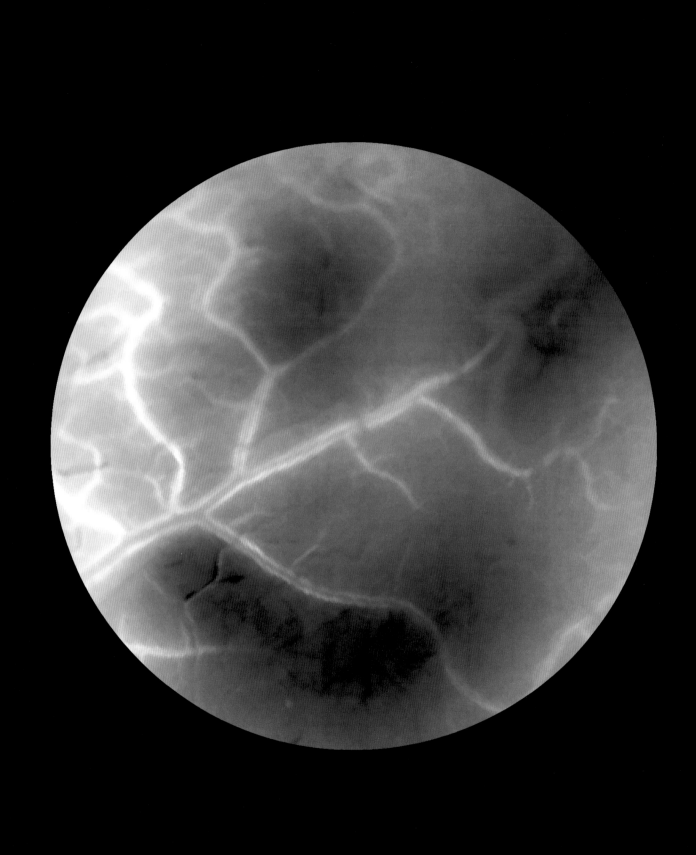

Intrinsic signal optical imaging.
Yevgeniy B. Sirotin and Aniruddha Das, 2006.

The link between blood volume and brain activity has been known for some time, although its precise nature remains mysterious, fueling ongoing debates in the field. What is certain is that at some level the amount of oxygenated blood delivered to neurons broadly correlates with their activity, presumably because they require extra energy when engaged. This activity can then be indirectly inferred by measuring blood volume. The first attempt to do so dates back to the late nineteenth century, when the Italian physiologist Angelo Mosso studied a patient, one L. Cane, whose brain was partially exposed in the back of his head because of a defect in his skull. In an experiment that might have failed to pass today's research-ethics boards, Mosso directly measured changes in the volume of his subject's pulsating brain as it responded to various stimuli. One in particular occasioned a distinct reaction: "I said just a few words expressing the impression that his wife had made upon me when I first saw her. Cane did not speak. The blood to his brain increased immediately and the [blood] volume of his feet markedly diminished."[6]

The ghostly image of a smiley face seen here was obtained using a method called intrinsic signal optical imaging, in which a high-speed video camera captured minute variations of local blood volume in a monkey's visual cortex (about ten millimeters across) as the animal stared at an illustration of a smiley face. The path from a yellow smiley face to the sophisticated neural events that gave rise to this astonishing snapshot is elaborate, yet occurs in a fraction of a second: The monkey's retina detects patterns of light and dark created by the illustration's eyes and mouth; the retina's output neurons relay this information to the thalamus deep inside the brain; the thalamus in turn relays this information to the visual cortex all the way in the back of the head, giving rise to local blood volume patterns in the cortex that correspond to the ones in the illustration. (Some of the more prominent blood-laden arteries are clearly visible in white.)

This image illustrates an important property of the visual cortex: It is laid out in such a way that neighboring objects in the world find themselves represented in neighboring areas in the brain. This ensures the faithful transmission of visual patterns observed in the world, like this smiley face, to the back of the brain. Of course, even if smiley faces do not land research grants, this form of imaging has proved to be an invaluable tool in studying the properties of the neocortex.

Orientation columns.

Yevgeniy B. Sirotin and Aniruddha Das, 2006.

This image also shows data obtained using intrinsic signal optical imaging—in fact, it was recorded in exactly the same area of visual cortex as the data shown on the previous spread—but focuses on an entirely different phenomenon. Neurons in the visual cortex excel at extracting lines from visual scenes, allowing us to quickly and efficiently recognize their basic features. When one looks at a desk, one does not experience it as a pointillist, pixel-like collection of dots, but rather as a clean shape defined by lines. Lines are the bread and butter of our visual experience: They define trees, horizons, the edges of things we don't want to bump into. Our visual system is designed to rapidly extract this meaningful information in order to make sense of the world. Consequently, the area in the visual cortex that first processes information coming in from the eyes is configured in a manner that reflects this preference for lines.

This data was obtained by filming a monkey's visual cortex while it looked at single lines drawn at different angles: some vertical, others horizontal, and yet others in between. It is rendered such that each pixel in the image represents the activity of a few hundred neurons and each color represents the angle of line that produced the strongest activity in the cortex. Thus, green clusters of neurons can be said to "prefer" vertical lines, whereas yellow ones are most responsive to horizontal ones. Together they underlie our exquisite sensitivity to the lines we routinely encounter in the world.

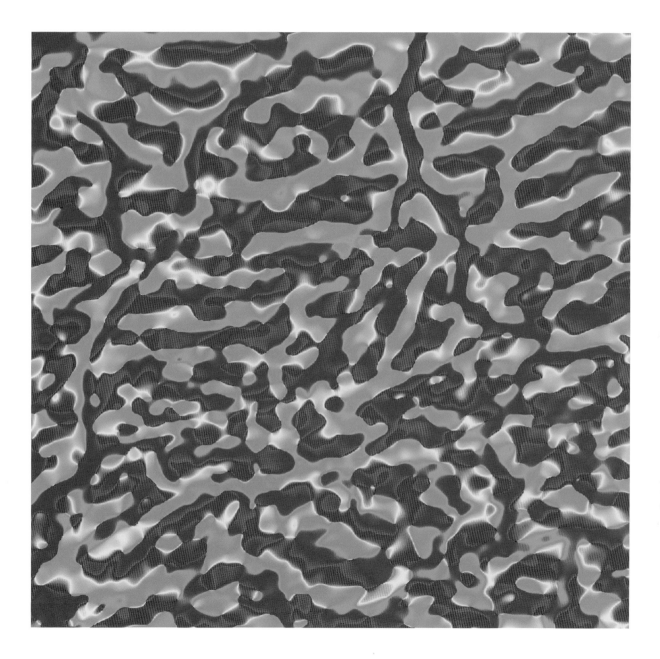

FUNCTIONAL MRI

Right: Basic organization of the human visual cortex.
Jack Gallant, 2009.

Overleaf: Category selectivity in the visual cortex.
Jack Gallant, 2007.

Like intrinsic signal optical imaging, fMRI measures the activity of neurons in the brain by detecting local changes in blood volume and oxygenation. fMRI, which can be performed in a regular MRI machine, takes advantage of the fact that oxygenated and deoxygenated blood exhibit different magnetic properties (due to the fact that the blood's hemoglobin contains a lot of iron). These magnetic properties locally affect the relaxation rates of protons in characteristic and recognizable ways. The tool presents the terrific advantage of being noninvasive and therefore has proven to be an invaluable method for studying human brain activity. The interpretation of fMRI data is far from straightforward, however, and remains an area of active research.

As for intrinsic signal optical imaging, fMRI does not measure neural activity directly. The connection between neural activity and the dynamics of blood volume is complex, and no matter how well we ultimately understand this relationship, measuring the latter will always provide an imperfect account of the former. Moreover, fMRI's relatively low resolution entails that each three-dimensional pixel in a data set represents not one but the combined activity of millions of neurons that may perform different functions. Time resolution is also an issue: Blood-volume dynamics occur over the course of seconds, whereas neural activity is hundreds of times faster, on a millisecond scale. Thus, it is fair to say that our perspective on neural activity recorded in fMRI experiments is quite limited. Yet, despite these confines, the tool has permitted unprecedented and in some cases positively astonishing access to the thinking human brain.

The fMRI data shown opposite reveals that the human visual cortex shares the same basic functional organization as the monkey's (see page 221): Neighboring points in a visual scene are mapped to neighboring points on the cortex. The data was obtained by recording a subject's brain response while he or she stared at a pattern of rings expanding out from the center of the gaze. Using a state-of-the-art data analysis algorithm, it was then possible to finely match up the distance of the ring relative to center, with activity recorded in the cortex. This image summarizes that relationship by color-coding an anatomical model of a partially inflated and unfolded view of the visual cortex: small rings at the very center of the gaze (in purples and reds) elicit a response in the center of the visual cortex (shown at right); as the ring pattern expands outward from the center (yellows, greens, and then blues), it activates cortical locations progressively further from the center of the visual cortex.

The data shown on the following spread illustrates a surprising finding. We are looking again at the visual cortex, this time for both hemispheres of the brain and in a flattened representation (much like a flattened map of the Earth). The black lines delineate known subdivisions of the visual cortex that specialize in different tasks (such as detecting straight lines or motion). What is remarkable is that parts of these regions are also found to perform some kind of basic categorization of visual scenes— a role one might normally expect to be carried out at a "higher level" in the brain, rather than at the entry point of raw sensory data flowing in from the eyes. The patches of red and blue represent areas that were activated by images of animate beings (portraits, full-body images, land and water animals), yellow and green by inanimate objects (indoor scenes, land and water panoramas, vehicles, textures, and text). Yellow and red regions were responsive specifically to views of large crowds, whereas green and blue areas were selective for textures. With new, advanced statistical techniques like the ones that were employed to analyze the fMRI data shown here, researchers can move beyond assigning gross functions to large areas of the cerebral cortex and instead map its activity at ever finer levels of detail.

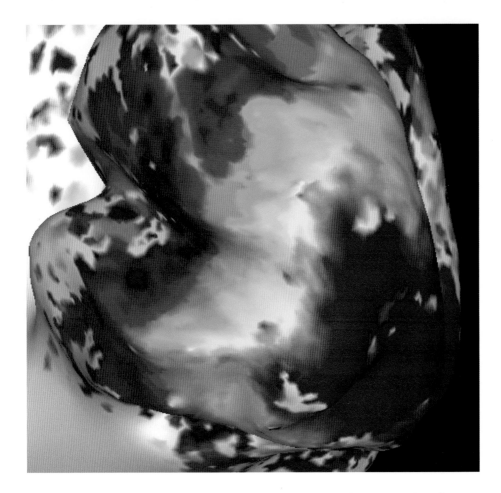

Functional MRI is a fledgling method that underwrites a fledgling science, and, like phrenology did generations ago, it promises to reveal the phenomena that underlie the intimate details of our mental lives. As such, the power of fMRI has inspired serious-minded researchers committed to applying the scientific method, but it has also been put to use in questionable studies positing oversize claims about the conclusions that can be drawn from fMRI studies. For beyond its methodological challenges lies a host of difficult philosophical issues, which come out of the woodwork when fMRI experiments seek to address complex questions about the mind. For instance, what does it mean, precisely, to assign a particular emotion to a small patch of gray matter when it is found to "light up" under particular circumstances? Does it make sense to think of one, or a small number, of brain regions as underlying complex thoughts, feelings, and behaviors? These are wide open questions, the solutions to which will not only provide a firmer grounding for fMRI studies, but will also yield profound insight into the relationship between mind and brain.

When the inconsistencies of phrenology were understood and a backlash ensued, we risked losing its positive contributions in the scuffle—most notably the theory that different parts of the cerebral cortex serve different functions, which remains a fundamental paradigm today. With fMRI we must also be careful not to throw out the baby with the bathwater, while at the same time maintaining a rigorous adherence to scientific standards, for in matters of the brain, a healthy dose of hard-nosed skepticism and blind optimism are called for in equal measure. After Galen and two millenia of fragmentary portraiture of the mind, this story is only just getting started.

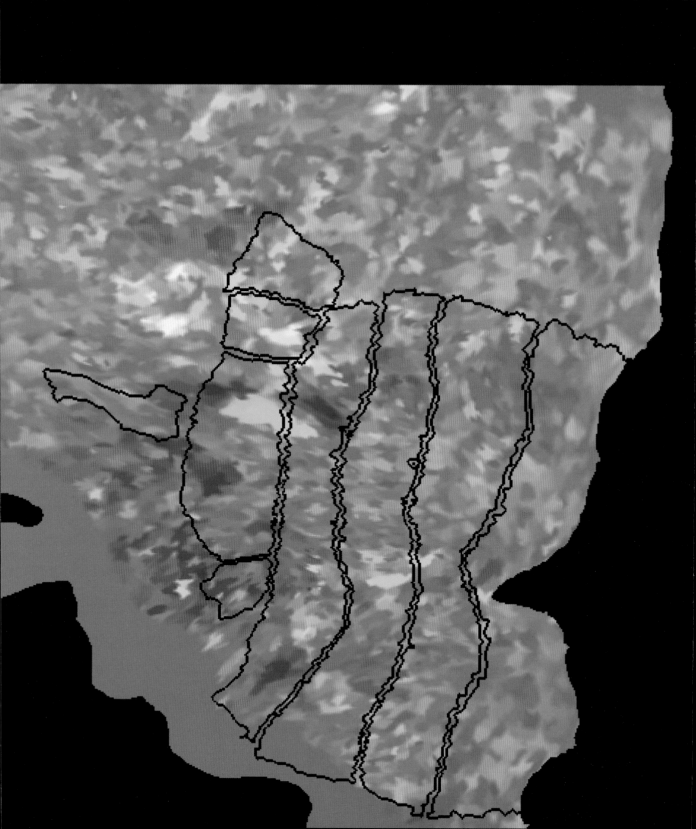

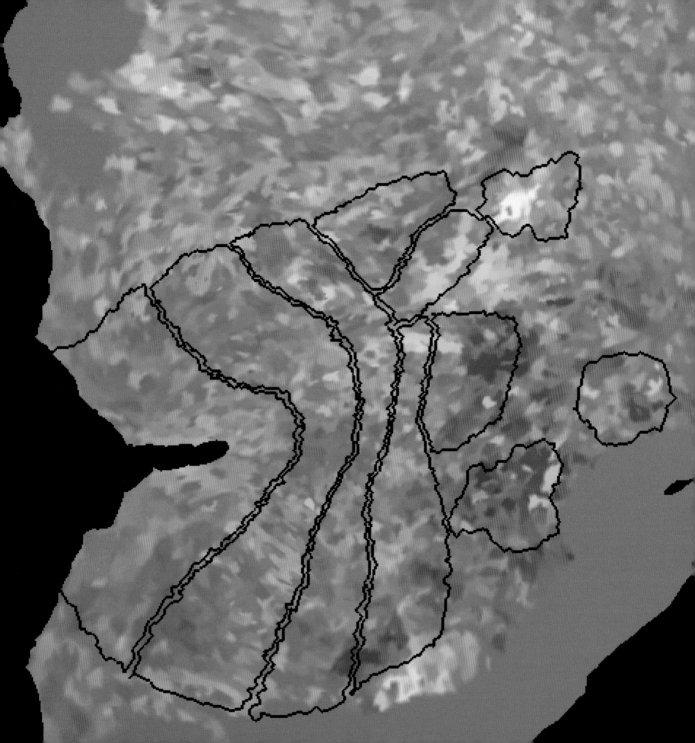

ENDNOTES

Preface

1. Lewitt, S. *Paragraphs on Conceptual Art.* *Artforum,* June 1967.

Chapter 1

1. Da Vinci, L. *A Treatise of Painting. Translated from the Original Italian; to which is Prefix'd, the Author's Life.* London: J. Senex and W. Taylor, 1721.

2. Stirling, W. *Some Apostles of Physiology.* London: Waterlow and Sons Ltd., 1902: 2.

3. O'Malley, C. D. O. *Andreas Vesalius of Brussels, 1514–1564.* Berkeley: University of California Press, 1964.

4. Ibid.

5. Willis, T. *The Anatomy of the Brain and Nerves.* Feindel, W., ed. Montreal: McGill University Press, 1965.

Chapter 2

1. Ramón y Cajal, S. *Recollections of My Life,* vol. 2, *The Story of my Scientific Work.* Madrid: Imprenta y Librería de Nicolás Moya, 1917. Translation from the Spanish (*Recuerdos de mi vida,* vol. 2. *Historia de mi labor científica*).

2. Ramón y Cajal, S. *Textura del sistema nervioso del hombre y de los vertebrados.* Madrid: Moya, 1899, 1904. Translation from the Spanish. Two English translations of the work are available in print: *Histology of the nervous system of man and vertebrates* (translated by Swanson, N., and L. W. Swanson). New York: Oxford University Press, 1995; and *Texture of the nervous system of man and the vertebrates* (an annotated and edited translation of the original Spanish text with the additions of the French version by P. Pasik and T. Pasik). New York: Springer Wien, 2002.

3. DeFelipe, J., and E. G. Jones. *Cajal on the Cerebral Cortex.* New York: Oxford University Press, 1988.

4. van Gehuchten, A. *Le Névraxe: Livre Jubilaire dédié a M. A. van Gehuchten.* vol. 12–13. Louvain: A. Uystpruyst (Librairie Universitaire), 1913: 29–45.

5. Ramón y Cajal, S. *Recollections of My Life,* vol. 2, *The Story of my Scientific Work.* Madrid: Imprenta y Librería de Nicolás Moya, 1917. Translation from the Spanish (*Recuerdos de mi vida,* vol. 2. *Historia de mi labor científica*).

6. Letter from Camillo Golgi to Nicolò Manfredi, Abbiategrasso, 16 February 1873. Translation from the Italian by P. Mazzarello. (see Mazzarello, P. *Golgi. A Biography of the Founder of Modern Neuroscience.* New York: Oxford University Press, 2009).

Chapter 3

1. Crick, F. *What Mad Pursuit: A Personal View of Scientific Discovery.* New York: Basic Books, 1990.

Chapter 4

1. Ramón y Cajal, S. *Recollections of My Life.* Translated by E. H. Craigie, with the assistance of J. Cano. Philadelphia: Am. Phil. Soc., 1937. Reprint Cambridge, MA: MIT Press, 1989.

Chapter 5

1. Hubel, D. H., and T. N. Wiesel. *Brain and Visual Perception: The Story of a 25-Year Collaboration.* New York: Oxford University Press, 2004.

2. Ibid.

3. Galvani, L. *Commentary on the effects of electricity on muscular motion.* Translated from the Italian by M. B. Foley. Norwalk, CT: Burndy Library, 1953, c. 1954.

4. Macpherson, L. J., et al. "The pungency of garlic: activation of TRPA1 and TRPV1 in response to allicin." *Current Biology* 15, no. 10 (24 May 2005): 929–34.

Chapter 6

1. Mel, B. W. "In the brain, the model is the goal." *Nature Neuroscience* 3 (November 2000): 1183.

Chapter 7

1. Crick, F. *The Astonishing Hypothesis: The Scientific Search for the Soul.* New York: Simon & Schuster, 1994.

2. Ibid.

3. Ibid.

4. Ibid.

5. Ibid.

6. Mosso A. *La Temperatura del Cervello.* Milan: Fratelli Treves, 1894. Translated from the Italian by C. Iadecola, in "Neurovascular regulation in the normal brain and in Alzheimer's disease." *Nature Reviews Neuroscience* 5 (May 2004): 348.

BIBLIOGRAPHY

Churchland, P., and T. J. Sejnowski. *The Computational Brain*. Cambridge, MA: MIT Press, 1994.

Clarke, E., and K. Dewhurst. *An Illustrated History of Brain Function: Imaging the Brain from Antiquity to the Present*. 2d ed. Novato, CA: Norman Publishing, 1996.

DeFelipe, J. *Cajal's Butterflies of the Soul: Science and Art*. New York: Oxford University Press, 2009.

Finger, S. *Origins of Neuroscience: A History of Explorations into Brain Function*. New York: Oxford University Press, 2001.

———. *Minds behind the Brain: A History of the Pioneers and Their Discoveries*. New York: Oxford University Press, 2004.

Golgi, C., and S. Ramón y Cajal. Nobel Lectures, 1906. Available online on the Nobel Foundation website: http://nobelprize.org/nobel_prizes/medicine/laureates/1906/

Hubel, D. H., and T. N. Wiesel. *Brain and Visual Perception: The Story of a 25-Year Collaboration*. New York: Oxford University Press, 2004.

Marshall, L. H., and H. W. Magoun. *Discoveries in the Human Brain: Neuroscience Prehistory, Brain Structure, and Function*. New York: Humana Press, 1998.

Martensen, R. L. *The Brain Takes Shape: An Early History*. New York: Oxford University Press, 2004.

Mazzarello, P. *Golgi: A Biography of the Founder of Modern Neuroscience*. New York: Oxford University Press, 2009.

McHenry, L. C., Jr., ed. *Garrison's History of Neurology*. Rev. ed. Springfield, IL: Charles C. Thomas Publishers Ltd., 1969.

Ramón y Cajal, S. *Recollections of My Life*. Translated by E. H. Craigie and J. Cano. Cambridge, MA: MIT Press, 1989.

———. *Cajal's Degeneration and Regeneration of the Nervous System*. Translated by R. M. May. Edited by J. DeFelipe and E. G. Jones. New York: Oxford University Press, 2009.

Wade, N. J. *A Natural History of Vision*. Cambridge, MA: MIT Press, 2000.

Zimmer, C. *Soul Made Flesh: The Discovery of the Brain—and How It Changed the World*. New York: Free Press, 2004.

SUGGESTED READING

Broks, P. *Into the Silent Land: Travels in Neuropsychology*. Conshohocken, PA: Atlantic Books, 2004.

Burrell, B. *Postcards from the Brain Museum: The Improbable Search for Meaning in the Matter of Famous Minds*. New York: Broadway Books, 2005.

Damasio, A. *Looking for Spinoza: Joy, Sorrow, and the Feeling Brain*. New York: Mariner Books, 2003.

———. *Descartes' Error: Emotion, Reason, and the Human Brain*. New York: Penguin, 2005.

Dennett, D. C. *Consciousness Explained*. New York: Penguin, 1993.

Kandel, E. R. *In Search of Memory: The Emergence of a New Science of Mind*. New York: W.W. Norton & Company, 2007.

Lehrer, J. *Proust Was a Neuroscientist*. Boston: Houghton Mifflin Harcourt, 2007.

———. *How We Decide*. Boston: Houghton Mifflin Harcourt, 2009.

Linden, D. *The Accidental Mind: How Brain Evolution Has Given Us Love, Memory, Dreams, and God*. Cambridge, MA: Belknap Press of Harvard University Press, 2008.

Livingstone, M. S. *Vision and Art: The Biology of Seeing*. New York: Abrams, 2008.

Ramachandran, V. S., and S. Blakeslee. *Phantoms in the Brain: Probing the Mysteries of the Human Mind*. New York: Harper Perennial, 1999.

Ramón y Cajal, S. *Advice for a Young Investigator*. Translated by N. Swanson and L. W. Swanson. Cambridge, MA: MIT Press, 2004.

Sacks, O. *Musicophilia: Tales of Music and the Brain*. New York: Picador USA, 2008.

———. *The Man Who Mistook His Wife for a Hat*. New York: Pan Books Ltd., 2009.

PICTURE CREDITS

tarius. Bologna: Ex Typographia Instituti Scientiarum, 1791. Photograph by Dwight Primiano.

Courtesy of Christine Constantinople: Pages 140–1, 151.

Courtesy of Thomas Deerinck and Mark Ellisman: Jacket front and pages 1, 74–5, 100–3, 105, 114–5, 124–5, 132–4.

Courtesy of Ryan W. Draft: Page 90. The Brainbow mouse was produced by J. Livet, T. A. Weissman, H. Kang, R. W. Draft, J. Lu, R. A. Bennis, J. R. Sanes, J. W. Lichtman, *Nature*, 2007, no. 450: 56–62.

Courtesy of Denis Ducreux: Page 212. Reprinted from Ozanne A., et al., *AJNR Am J Neuroradiol* 28, no. 7 (Aug 2007): 1271–9. Copyright 2007, with permission from the American Society of Neuroradiology.

Courtesy of Andy Fischer: Jacket back and page 180.

Courtesy of Jack Gallant: Pages 198–9, 225–7.

Pages 198–9, 225: See Hansen, K. A., et al., *J. Neurosci* 27, no. 44 (31 Oct 2007): 11896–911.

Courtesy of Alexander Gottschalk: Page 164. See Liewald, J. F., et al., *Nat Methods* 5, no. 10 (Oct 2008): 895–902.

Courtesy of Patric Hagmann: Page 210. See Hagmann, P., et al., *J. Neurosci Methods*, 22 Jan 2010.

Reprinted under the Creative Commons Attribution License from Hagmann, P., et al., *PLoS Biol* 6, no. 7 (1 Jul 2008): e159: Page 214.

Courtesy of Dr. Raqeeb Haque, MD: Pages 206–7, 209. Photograph by Daniel Jordan.

Courtesy of Kristen M. Harris: Pages 122–3. Reprinted from K. M. Harris and D. M. Landis, *Neuroscience* 19, no. 3 (Nov 1986): 857–72. Copyright 1986, with permission from Elsevier.

Courtesy of Stefan Hell, MPI Biophysical Chemistry, Göttingen: Pages 138–9.

Courtesy of Michael Hendricks: Pages 4, 98, 182–3.

Pages 4, 98: See Hendricks, M., and S. Jesuthasan, *J. Neurosci* 29, no. 20 (20 May 2009): 6593–8.

Image produced by John Heuser, MD: Pages 127–8.

Photograph by Justin Keena: Page 165.

Courtesy of In-Jung Kim: Endpaper (front) and pages 84–5. See Kim, I. J., et al., *Nature* 452, no. 7186 (27 Mar 2008): 478–82.

Image by Bernd Knöll and Jürgen Berger: Page 126. With permission from Society for Neuroscience. See Stern, S., et al., *J. Neurosci* 29, no. 14 (8 Apr 2009): 4512–8.

Courtesy of Clay Lacefield: Pages 153–5, 158–9.

Courtesy of Jean Livet: Binding case and pages 92–3. The Brainbow mouse was produced by J. Livet, T. A. Weissman, H. Kang, R. W. Draft, J. Lu, R. A. Bennis, J. R. Sanes, J. W. Lichtman, *Nature*, 2007, no. 450: 56–62.

Courtesy of David Lyon: Page 113. See Wickersham, I. R., et al., *Neuron* 53, no. 5 (1 Mar 2007): 639–47.

Courtesy of Henry Markram, Founder & Director, Blue Brain Project: Pages 166–7, 192. See Markram, H., *Nat Rev Neurosci* 7, no. 2 (Feb 2006): 153–60.

Courtesy of Roderick MacKinnon: Pages 136–7. See Long, S. B., et al., *Science* 309, no. 5736 (5 Aug 2005): 903–8.

Courtesy of Dr. Paolo Mazzarello, University of Pavia—University Museum System—Museum for the History of University: Pages 12–3, 47, 58.

Pages 12–3, 47: See Golgi, C., "Sulla fina struttura dei bulbi olfattori," in *Rivista Sperimentale di Freniatria e Medicina Legale*, 1875, vol. 1: 403–25.

Page 58: Letter from Camillo Golgi to Nicolò Manfredi, Abbiategrasso, February 16, 1873.

Courtesy of Kieran Murphy: Page 215.

Courtesy of Albert Pan: Pages 2–3. See Pan, Y. A., et al, "Multicolor Brainbow imaging in larval zebrafish," in *Imaging in Developmental Biology: A Laboratory Manual*, eds. Wong, R., J. Sharpe, and R. Yuste, Cold Spring Harbor Laboratory Press (in press).

Courtesy of Emmanuel Procyk: Page 162. Reprinted from Quilodran, R., et al., *Neuron* 57, no. 2 (24 Jan 2008): 314–25. Copyright 2008, with permission from Elsevier.

Courtesy of Alfonso Rodríguez-Baeza: Page 216.

The Royal Collection © 2009 Her Majesty Queen Elizabeth II. [Shelfmark RCIN 919127r]: Pages 26–7.

Courtesy of Alexandre Saez: Page 157. See Paton, J. J., et al., *Nature* 439, no. 7078 (16 Feb 2006): 865–70.

Courtesy of Sanes Laboratory: Page 81. See Lichtman, J. W., et al., *Nature Reviews Neuroscience* 9, no. 6 (Jun 2008): 417–22.

Courtesy of Nathaniel Sawtell: Page 112.

Courtesy of Carl Schoonover: Pages 118–9.

Courtesy of Sebastian Seung: Page 189. See Turaga, S. C., et al., *Neural Comput* 22, no. 2 (Feb 2010): 511–38.

Courtesy of Nathan Shaner: Page 89. See Shaner N. C., et al., *Nat Methods* 2, no. 12 (Dec 2005): 905–9.

Courtesy of Yevgeniy B. Sirotin: Pages 220, 223. See Sirotin, Y. B., and A. Das, *Nature* 457, no. 7228 (22 Jan 2009): 475–9; and Sirotin, Y. B., et al., *Proc Natl Acad Sci USA* 106, no. 43 (27 Oct 2009): 18390–5.

Image by Josef Spacek: Page 130. Courtesy of Synapse Web Atlas for Ultrastructural Neurocytology, http://synapses.clm.utexas.edu/.

Courtesy of Greg Stuart: Page 152. See Stuart, G. J., and B. Sakmann, *Nature* 367, no. 6458 (6 Jan 1994): 69–72.

Courtesy of the Süleymaniye Library (Istanbul): Page 18. See Al-Haytham, I., "Book of Optics" (1087).

Courtesy of Gulsen Surmeli: Page 86. See Tanabe, Y., et al., *Cell* 95, no. 1 (2 Oct 1998): 67–80.

Courtesy of Edgar Toro: Pages 160–1.

Courtesy of Lav Varshney: Page 186. See Varshney, L. R., et al., 28 Oct 2009: arXiv:0907.2373v2.

Courtesy of Henning U. Voss: Page 213.

Courtesy of Tamily A. Weissman: Pages 94–5, 172, 177.

Pages 94–5, 177: The Brainbow mouse was produced by J. Livet, T. A. Weissman, H. Kang, R. W. Draft, J. Lu, R. A. Bennis, J. R. Sanes, J. W. Lichtman, *Nature*, no. 450 (2007): 56–62.

Page 172: Triple transgenic cross by Judy Tollett. See Feng, G., et al., *Neuron* 28, no. 1 (Oct 2000): 41–51.

Wellcome Library, London: Pages 25, 29–30, 32–4, 37.

Page 34: See Descartes, R., *L'homme*, Charles Angot, Paris, 1664.

Page 29: See Estienne, C., *De dissectione partium corporis humani libri tres …/Una cum figuris et incisionum declarationibus, a Stephano Riverio chirurgo composites*, S. Colinaeus, Paris, 1545.

Page 25: See Fludd, R., *Tomi secundi tractatus primi, sectio secunda, de technica microcosmi historia. Tomi secundi tractatus secundus, de praeternaturali utriusque mundi historia*, Frankfurt, E. Kempffer, for J. T. de Bry, 1619–21.

Pages 30, 32–3: See Vesalius, A., *De humani corporis fabrica libri septem*, Joannem Oporinum, Basel, 1555.

Page 37: See Willis, T., *Cerebri anatome: cui accessit nervorum descriptio et usus*, G. Schagen, Amsterdam, 1664.

Courtesy of Lasani Wijetunge: Endpaper (back) and pages 106–8.

Pages 106–8: See Wijetunge, L. S., et al., *J. Neurosci* 28, no. 49 (3 Dec 2008): 13028–37.

Courtesy of Guang Yang: Page 120. See Yang, G., et al., *Nature* 462, no. 7275 (17 Dec 2009): 920–4.

Courtesy of Malcolm Young: Page 184. See Young, M. P., et al., *Rev. Neurosci* 5, no. 3 (Jul–Sep 1994): 227–50.

INDEX

Page numbers in *italics* indicate
photographs and illustrations.

ACKNOWLEDGMENTS

I thank the many research scientists around the world whose work is featured in these pages. Without their generosity, expertise, and advice this project would not have been possible. I am particularly indebted to the eight senior scientists—Nicholas Wade, Javier DeFelipe, Joshua Sanes, Michael Goldberg, Maryann Martone, Mark Ellisman, Terrence Sejnowski, and Joy Hirsch—who contributed the chapters' opening essays, lending a diversity of voices, and their collective experience with the themes and methods presented in the book.

I am thankful for the early conversations I had with Larry Abbott, Amelia Atlas, Tom Deerinck, Madeleine Elish, Mallory Jensen, Jayant Kulkarni, Clay Lacefield, Gül Russell, and Greg Wayne about how to organize and present the content. I am grateful in particular for the invaluable advice I received from Josh Sanes and Javier DeFelipe during early planning stages and throughout the project.

In order to ensure that the facts are presented accurately, I appealed to the collective expertise of those who contributed the images and also submitted the text to friends and colleagues for critiquing. Thanks in particular go to Andrew Fink, Stuart Firestein, and Tim Requarth—a.k.a. the Death Panel—for combing through my prose several times over. I am grateful also for the watchful eyes of Joe Barfett, Aniruddha Das, Paolo Mazzarello, Maxim Nikitchenko, and Zev Rosen. Any mistakes that slipped through the cracks are my responsibility alone and will never cease to haunt me. I received invaluable feedback from non-scientists, including Madeleine Elish, Amanda Freeman, and Alida Lasker about the prose and its accessibility. I am particularly grateful for Meehan Crist's keen editorial eye and for Karan Mahajan's heroic read-through of a draft so preliminary it probably should never have been shown to anyone literate.

I received outstanding bibliographical and reference assistance from Patricia Buckingham (Bodleian Library, Oxford); Consuelo Dutschke (Columbia University Rare Books Library, New York); Carla Garbarino (University of Pavia); Gisselle Garcia (American Museum of Natural History, New York); Katie Holyoak (Royal Collection, London); Emma Jacobs, Russell Johnson (UCLA Neuroscience History Archives, Los Angeles); Stephen Novak (Columbia University Health Sciences Special Collections Library, New York); Janet Parks (Columbia University Fine Arts Library, New York); Arlene Shaner (New York Academy of Medicine, New York); Anna Smith (Wellcome Trust, London); Benedek Varga (Semmelweis Museum, Budapest); and the management of the Süleymaniye Library in Istanbul.

For their assistance in locating and obtaining images, I am deeply indebted to Thania Benios, Christiane Debono, Phyllis Kisloff, John Martin, Kelly Overly, Patrick Parker, Francisco Rodriguez, Aelfie Tuff, Tamily Weissman, and especially Gulsen Surmeli and Elaine Zhang. I thank Eszter Blahák, Daniel Jordan, Justin Keena, and Dwight Primiano for their excellent photography work. I am very grateful for the kindness, generosity, and deep expertise of Miguel Freire, Virginia Marin, and especially Juan de Carlos of the Cajal Institute in Madrid.

I thank the scientific mentors who have nurtured my interest in neuroscience and introduced me to some of the techniques presented in the book: Jian Yang, Chris McBain, Hiroshi Nishimune, Josh Sanes, Randy Bruno, and Carol Mason. And I am indebted to the writers and editors who gave me the encouragement and confidence I needed to undertake and complete the project: Stéphane Marchand, Susan Fraker, Karan Mahajan, and especially Elaine Sciolino.

The germ of this book can be found in conversations I enjoyed early on in this project with the founding members of NeuWrite, a science-writing workshop associated with Columbia University. I thank Thania Benios, Meehan Crist, Kate Daloz,

Anthony DeCostanzo, Jayant Kilkarni, Clay Lacefield, Cindy Lang, Michelle Legro, Abigail Rabinowitz, David Schneider, Kim Tingley, Greg Wayne, and especially our hosts Stuart Firestein and Diana Reiss, for their advice, support, and terrific company. Stuart's clarion call for us to develop strategies for writing about the methods sections of scientific publications was a significant source of inspiration.

I am deeply grateful to my editor at Abrams, Andrea Danese, who first approached me with the idea of a visual book based on images of the brain gathered from the latest research in neuroscience, and who tirelessly helped me carry this project from proposal to publication. I thank Caitlin Kenney, editorial assistant at Abrams, as well as Gary Tooth of Empire Design Studio for his compelling layout.

The book benefited substantially from my close friendship with my two housemates Tim Requarth and Andrew Fink, invaluable sparring partners whose saintliness I became firmly convinced of as they reread the manuscript ad nauseam and humored my antics as deadlines loomed. Thank you to my parents, John and Susan, as well as my siblings Daniel and Julie, for their encouragement throughout the process. I owe my deepest expression of gratitude to Samantha Holmes for her unwavering support and her sure, incisive taste in matters visual and writerly. It was a relief to know that when in doubt I could always trust in her judgment; without her mark, the final product would undoubtedly have been far inferior.

I dedicate this book to Monsieur Ferraris, who taught me many years ago that biology can be beautiful, even when it is confusing.

Page 1: Glial cell imaged with a scanning electron microscope. Thomas Deerinck and Mark Ellisman, 2009.

Pages 2–3: Brainbow zebrafish larva. Albert Pan, 2008.

Page 4: Scaffolding in axons (see page 99).

Endpaper (front): JAM-B cells in the mouse retina (see page 82).
Endpaper (back): Antibody staining of a mouse cerebral cortex (see page 104).

Editor: Andrea Danese
Designer: Gary Tooth, Empire Design Studio
Production Manager: Ankur Ghosh

Library of Congress Cataloging-in-Publication Data

Schoonover, Carl E.
    Portraits of the mind : visualizing the brain from antiquity to the 21st century / Carl E. Schoonover.
        p. cm.
    Includes bibliographical references and index.
    ISBN 978-0-8109-9033-3 (alk. paper)
    1. Brain—Anatomy—Atlases. 2. Brain—Imaging. I. Title.
    QM455.S35 2010
    612.8'2—dc22

            2010005999

Published in 2010 by Abrams, an imprint of ABRAMS. All rights reserved. No portion of this book may be reproduced, stored in a retrieval system, or transmitted in any form or by any means, mechanical, electronic, photocopying, recording, or otherwise, without written permission from the publisher.

Printed in the U.S.A.
10 9 8 7 6 5 4 3 2

Abrams books are available at special discounts when purchased in quantity for premiums and promotions as well as fundraising or educational use. Special editions can also be created to specification. For details, contact specialmarkets@abramsbooks.com, or the address below.

ABRAMS
THE ART OF BOOKS SINCE 1949

115 West 18th Street
New York, NY 10011
www.abramsbooks.com

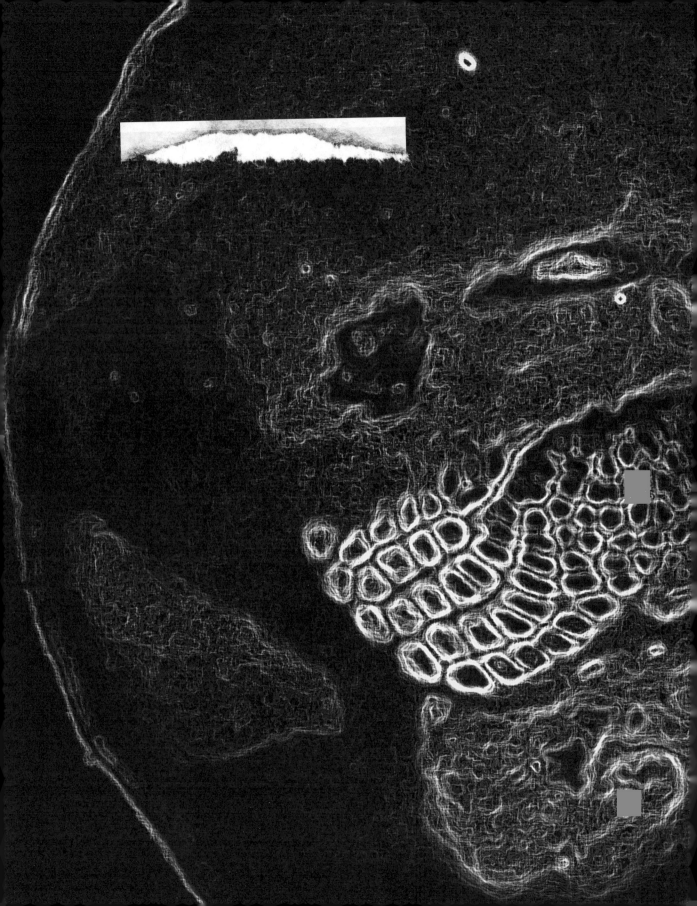